Energy Science, Engineering and Technology

Modular Electricity Storage: Benefits and Costs

ENERGY SCIENCE, ENGINEERING AND TECHNOLOGY

Additional books in this series can be found on Nova's website under the Series tab.

Additional E-books in this series can be found on Nova's website under the E-books tab.

ELECTRICAL ENGINEERING DEVELOPMENTS

Additional books in this series can be found on Nova's website under the Series tab.

Additional E-books in this series can be found on Nova's website under the E-books tab.

ENERGY SCIENCE, ENGINEERING AND TECHNOLOGY

MODULAR ELECTRICITY STORAGE: BENEFITS AND COSTS

BRENT N. MENDELL
AND
LISA P. BRUNWICK
EDITORS

Nova Science Publishers, Inc.
New York

Copyright © 2012 by Nova Science Publishers, Inc.

All rights reserved. No part of this book may be reproduced, stored in a retrieval system or transmitted in any form or by any means: electronic, electrostatic, magnetic, tape, mechanical photocopying, recording or otherwise without the written permission of the Publisher.

For permission to use material from this book please contact us:
Telephone 631-231-7269; Fax 631-231-8175
Web Site: http://www.novapublishers.com

NOTICE TO THE READER

The Publisher has taken reasonable care in the preparation of this book, but makes no expressed or implied warranty of any kind and assumes no responsibility for any errors or omissions. No liability is assumed for incidental or consequential damages in connection with or arising out of information contained in this book. The Publisher shall not be liable for any special, consequential, or exemplary damages resulting, in whole or in part, from the readers' use of, or reliance upon, this material. Any parts of this book based on government reports are so indicated and copyright is claimed for those parts to the extent applicable to compilations of such works.

Independent verification should be sought for any data, advice or recommendations contained in this book. In addition, no responsibility is assumed by the publisher for any injury and/or damage to persons or property arising from any methods, products, instructions, ideas or otherwise contained in this publication.

This publication is designed to provide accurate and authoritative information with regard to the subject matter covered herein. It is sold with the clear understanding that the Publisher is not engaged in rendering legal or any other professional services. If legal or any other expert assistance is required, the services of a competent person should be sought. FROM A DECLARATION OF PARTICIPANTS JOINTLY ADOPTED BY A COMMITTEE OF THE AMERICAN BAR ASSOCIATION AND A COMMITTEE OF PUBLISHERS.

Additional color graphics may be available in the e-book version of this book.

Library of Congress Cataloging-in-Publication Data

Modular electricity storage : benefits and costs / [edited by] Brent N. Mendell and Lisa P. Brunwick.
    p. cm.
    Includes bibliographical references and index.
    ISBN 978-1-61470-459-1 (hardcover : alk. paper) 1. Energy storage. 2. Modularity (Engineering) I. Mendell, Brent N. II. Brunwick, Lisa P.
    TK2980.M63 2011
    333.793'2--dc23
                       2011023238

Published by Nova Science Publishers, Inc. ✝ *New York*

CONTENTS

Preface		vii
Chapter 1	Electric Utility Transmission and Distribution Upgrade Deferral Benefits from Modular Electricity Storage *Jim Eyer*	1
Chapter 2	Benefit/Cost Framework for Evaluating Modular Energy Storage *Susan M. Schoenung and Jim Eyer*	79
Index		109

PREFACE

An expanding array of state-of-the-art and emerging technologies are opening new opportunities to improve and/or to reduce the costs to generate, deliver and use electricity. One of these technologies is modular electricity storage (MES). In this book, the specific utility application addressed is the use of MES to reduce the cost of electricity delivery by reducing the cost of electricity transmission and distribution (T&D) equipment. Specifically, MES would be used to defer expensive improvements or capacity additions to T&D equipment by providing modular capacity additions as needed.

Chapter 1 - An expanding array of state-of-the-art and emerging technologies are opening new opportunities to improve and/or to reduce the costs to generate, deliver, and use electricity. One of these technologies is addressed in this report: modular electricity storage (MES). The specific utility application addressed herein is the use of MES to reduce the cost of electricity delivery by reducing the cost of electricity transmission and distribution (T&D) equipment. Specifically, MES would be used to defer expensive improvements or capacity additions to T&D equipment by providing *modular* capacity additions *as needed*.

Chapter 2- The work documented in this report was undertaken for three key purposes: Often, benefits and costs developed in previous energy storage studies were computed using different financial bases. This work reconciles those financial bases so that costs and benefits are expressed using consistent bases and assumptions. The Energy Storage Systems (ESS) Program management at Sandia wanted to update their storage technology cost and performance information to reflect state-of-the-art. Results in this report reflect another next step in the ongoing investigation of innovative and potentially attractive value propositions for electricity storage by the Department of Energy (DOE) and Sandia National Laboratories (SNL) ESS Program.

In: Modular Electricity Storage
Eds: B.N. Mendell and L.P. Brunwick

ISBN: 978-1-61470-459-1
© 2012 Nova Science Publishers, Inc.

Chapter 1

ELECTRIC UTILITY TRANSMISSION AND DISTRIBUTION UPGRADE DEFERRAL BENEFITS FROM MODULAR ELECTRICITY STORAGE[*]

Jim Eyer

NOTICE

Prepared by
Sandia National Laboratories
Albuquerque, New Mexico 87185 and Livermore, California 94550
Sandia is a multiprogram laboratory operated by Sandia Corporation, a Lockheed Martin Company, for the United States Department of Energy's National Nuclear Security Administration under Contract DE-AC04-94AL85000.
Approved for public release; further dissemination unlimited.

Issued by Sandia National Laboratories, operated for the United States Department of Energy by Sandia Corporation.

---

[*] This is an edited, reformatted and augmented version of A Study for the DOE Energy Storage Systems Program publication, Sandia Report SAN D2009-4070, dated June 2009.

This report was prepared as an account of work sponsored by an agency of the United States Government. Neither the United States Government, nor any agency thereof, nor any of their employees, nor any of their contractors, subcontractors, or their employees, make any warranty, express or implied, or assume any legal liability or responsibility for the accuracy, completeness, or usefulness of any information, apparatus, product, or process disclosed, or represent that its use would not infringe privately owned rights. Reference herein to any specific commercial product, process, or service by trade name, trademark, manufacturer, or otherwise, does not necessarily constitute or imply its endorsement, recommendation, or favoring by the United States Government, any agency thereof, or any of their contractors or subcontractors. The views and opinions expressed herein do not necessarily state or reflect those of the United States Government, any agency thereof, or any of their contractors.

Printed in the United States of America. This report has been reproduced directly from the best available copy.

ABSTRACT

The work documented in this report was undertaken as part of an ongoing investigation of innovative and potentially attractive value propositions for electricity storage by the United States Department of Energy (DOE) and Sandia National Laboratories (SNL) Electricity Storage Systems (ES S) Program. This study characterizes one especially attractive value proposition for modular electricity storage (ME S): electric utility transmission and distribution (T&D) upgrade deferral. The T&D deferral benefit is characterized in detail. Also presented is a generalized framework for estimating the benefit. Other important and complementary (to T&D deferral) elements of possible value propositions involving MES are also characterized.

ACKNOWLEDGMENT

This work has been sponsored by the United States Department of Energy (DOE) Energy Storage Systems (ES S) Program under contract to Sandia National Laboratories (SNL). The author would like to thank Imre Gyuk of DOE for his support of this work, Paul Butler of SNL for his thorough and insightful review, and Nancy Clark of SNL for her invaluable assistance and support. Special thanks go to Garth Corey for his helpful and timely comments and suggestions.

SNL is a multiprogram laboratory operated by Sandia Corporation, a Lockheed Martin Company for the United States Department of Energy's National Nuclear Security Administration under Contract DE-AC04-94AL85000.

ACRONYMS AND ABBREVIATIONS

AC	alternating current
AEP	American Electric Power
DER	distributed energy resource
DG	distributed generation
DOE	United States Department of Energy
DR	demand response
DU	distributed utility
DUA	Distributed Utility Associates
DUPS	Distributed Utility Penetration Study
EOYB	end of year balance
EPRI	Electric Power Research Institute
ESS	Energy Storage Systems Program
FERC	Federal Energy Regulatory Commission
GENSET	engine/generator "set" (system)
ISO	independent system operator
kVA	kiloVolt Amp(s)
kW	kiloWatts
MVA	MegaVolt Amp(s)
MW	MegaWatts
MES	modular electricity storage
Na/S	sodium/sulfur
NGK	NGK Insulators, Ltd.
ORNL	Oak Ridge National Laboratories
PCU	power conditioning unit
PG&E	Pacific Gas and Electric Company
PJM	PJM Interconnection
PQ	power quality
PSI	Public Service of Indiana
PW	Present Worth

RAP	Rate Assistance Program
RTO	regional transmission organization
SCE	Southern California Edison
SDG&E	San Diego Gas and Electric
SNL	Sandia National Laboratories
T&D	transmission and distribution
THD	total harmonic distortion
UPS	uninterruptible power supply
VAR	Volt Amp(s) reactive
VRB	VRB Power Systems

EXECUTIVE SUMMARY

An expanding array of state-of-the-art and emerging technologies are opening new opportunities to improve and/or to reduce the costs to generate, deliver, and use electricity. One of these technologies is addressed in this report: modular electricity storage (MES).

The specific utility application addressed herein is the use of MES to reduce the cost of electricity delivery by reducing the cost of electricity transmission and distribution (T&D) equipment. Specifically, MES would be used to defer expensive improvements or capacity additions to T&D equipment by providing *modular* capacity additions *as needed*.

Criteria that indicate whether MES might be viable for T&D deferral include:

1) High T&D cost.
2) High peak-to-average demand ratio.
3) Modest projected overload.
4) Slow peak demand growth (rate).
5) Uncertainty about the timing and/or likelihood of block load additions.
6) T&D construction delays or construction resource constraints.
7) T&D upgrade competes with other important projects for capital.
8) The same MES system used for T&D deferral provides additional benefits – revenue or avoided cost – that can be aggregated into an attractive value proposition, such as on-peak energy and electric supply capacity.

MES is especially well-suited to those locations where air emissions regulations, noise regulations, fuel storage or other safety-related challenges restrict the use of combustion-based distributed *generation,* and/or where the price differential is large between times when storage is charged and when it is discharged.

A key result of the work described in this report is a generalized framework for estimating the financial benefit of deferring a T&D upgrade for one year with MES. The framework has three fundamental steps:

1) Estimate T&D cost.
2) Determine storage sizing – power and discharge duration.
3) Estimate T&D deferral benefit.

The first step is to estimate the installed cost for the T&D equipment to be deferred. That is the cost to design, purchase and install the T&D equipment. Typical values fall within the range of $25 to $250 per kW of T&D capacity installed.

The second step is to determine the amount (size) of storage needed to defer a specific T&D upgrade for the next year. The storage sizing evaluation is primarily based on two criteria: 1) storage power rating and 2) storage discharge duration (i.e., the amount of time that storage can be discharged, at its power rating).

For the purpose of this analysis, storage *power* rating is expressed in terms of the amount of storage capacity (kW) needed relative to the existing T&D capacity (storage power). Consider an example: for a 12 MW transformer, 3% storage power equates to storage whose power rating is 0.03 * 12 MW = 360 kW. The various plots in Figure ES-1, below, reflect a range of storage power values – ranging from storage rated at 1% of the existing T&D capacity to storage rated at 8% of the existing T&D capacity.

The final step is to estimate the single-year benefit for deferring a T&D upgrade whose cost is known. The single-year T&D deferral benefit is shown on the vertical axis of Figure ES-1. Benefit values are expressed in units of dollars per kW of storage needed (storage power).

Two other important criteria reflected in Figure ES-1 are: 1) T&D upgrade factor and 2) fixed charge rate. T&D upgrade factor reflects the amount of load carrying capacity to be *added* to the T&D system as part of the upgrade. Benefit values shown in ES-1 are for an upgrade factor of 0.33. (Example: equipment rated at 12 MW is replaced with equipment rated at 16 MW. The upgrade factor is 4/12 = 0.33.)

The fixed charge rate is used to estimate the *annual* cost of utility capital equipment based on the total *installed* cost of the equipment. An example: for a plant costing $1 Million, if using a fixed charge rate of 0.11, the annual cost for the plant is $110,000 ($1,000,000 * 0.11 = $110,000).

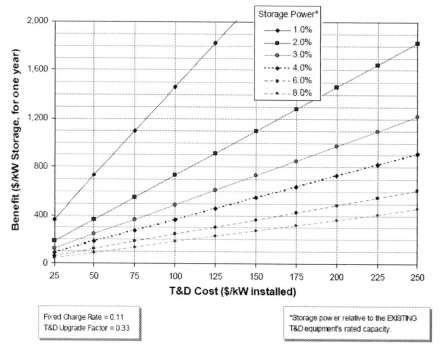

Figure ES-1. Annual (single year) Benefit for Storage Used for T&D Deferral.

Benefit values plotted in Figure ES-1 reflect a representative fixed charge rate of 0.11 (or 11% of the equipment's installed cost, annually).

Consider an example of how Figure ES-1 would be used. T&D equipment rated at 12 MW will be upgraded to 16 MW. To defer that upgrade for one year, power engineers specify storage whose power rating is 3% of the rating of the existing T&D equipment (that's 0.03 * 12 MW = 360 kW).

As shown in Figure ES-2, if the T&D equipment upgrade being installed costs about $100/kW, then the *single-year* benefit for deferring the upgrade is about $500/kW of storage (that is $500/kW of storage for deferring the upgrade *for one year*).

Note that while the generalized framework depicted in Figure ES-2 specifies *storage power,* it does include *implicit* consideration of discharge duration. That is, the storage system used must have both the power and discharge duration (energy storage) needed.

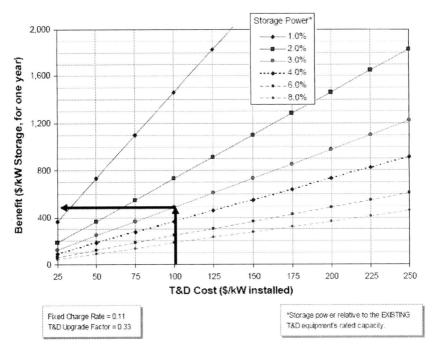

Figure ES-2. Annual (single year) Benefit for Storage for T&D Deferral, Example.

Both power and discharge duration are established during the storage sizing process. And, both are very circumstance-specific: discharge duration needs are driven mostly by the *load shape* during peak loading, whereas storage power rating is driven by the *magnitude of the load*. Typical values for discharge duration range from three hours to six hours.

Study results indicate circumstances for which incremental or modular capacity additions – including MES – may be the most attractive option to serve load additions and/or to accommodate normal load growth. This option is especially attractive if: a) there is uncertainty about the need for an upgrade or about availability of resources to complete an upgrade when needed, or b) the modular resources are transportable and readily redeployable. Such benefits could be a key element of attractive value propositions for many MES projects. Furthermore, aggregation of two or more individual benefits may be an important way to increase the number of opportunities for economically viable, grid-connected MES.

1. INTRODUCTION

This section describes utility transmission and distribution (T&D) deferral benefits that could be realized if modular distributed energy resources (DERs) are used to serve a portion of peak demand so that the utility may delay (defer) large T&D equipment upgrades (investments). DERs include modular electricity storage (MES), distributed generation (DG), geographically-targeted demand response (DR), and energy efficiency.

The DER concept is well-suited to the following circumstances:

1) Peak demand on a T&D node is at or near the T&D equipment's load carrying capacity (limit) – resulting in a "hot spot," and
2) A relatively small amount of DER capacity located downstream (electrically) from the hot spot can serve a portion of peak demand, on the margin, such that an upgrade of the T&D equipment is deferrable.

Modular DERs allow small capacity additions, on the margin, to defer a much larger lump investment in T&D equipment. The term "lump investment" refers to the fact that, by the nature of electric utility equipment, load carrying capacity cannot be added to the T&D system in small increments. Typical upgrade factor values range from 0.25 (add 25% more capacity) to as high as 0.50 or more.[1]

For the purpose of this report, T&D includes the distribution system and what is referred to as "subtransmission" (equipment/circuits between the transmission and distribution systems whose voltage is one of the following: 69 kV, 115 kV, or 138 kV).

Modular Electricity Storage for T&D Deferral, an Example

To illustrate the DER concept, it is helpful to consider an example involving MES: a specific node within a utility's distribution system whose rated load carrying capacity is 12 MW (12,000 kW). Peak demand in the current year (year 0) is 11.85 MW, 1.25% below the equipment's load carrying capacity, and it is growing at about 2%/year. Based on this load growth rate, peak demand in the next year (year 1) will exceed the 12 MW rating by about 87 kW, or 0.7%. To avoid overloading the equipment, engineers plan to increase load carrying capacity by 33%, to 16 MW (upgrade factor of 0.33). The case is shown graphically in Figure 1.

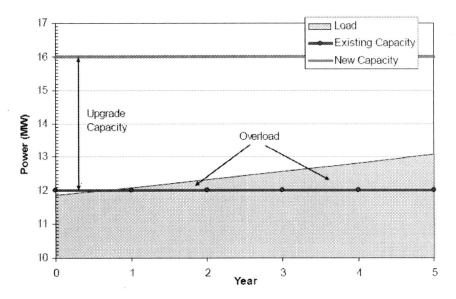

Figure 1. Load, Existing Capacity, and Upgraded Capacity.

For year 1, utility power engineers decide that 150 kW of electricity storage will serve sufficient load such that the utility may defer the upgrade. The 150 kW of storage is 72% above the projected overload of 87 kW, providing a margin of 63 kW additional storage capacity to allow for uncertainty.

In year 2, the forecasted overload increases to 329 kW. Power engineers then decide to use a storage system whose rating is 500 kW (52% above the projected overload) to defer the T&D upgrade for the second year. The 500 kW system is 52% above the projected overload of 329 kW, providing a margin of 171 kW additional storage capacity to allow for uncertainty.

Note that the values used for this example illustrate the concept of deferral; however, they do not include consideration of common phenomena that reduce the efficacy of power delivery, such as harmonic currents and reactive power. To account for effects related to such phenomena, T&D capacity and storage systems have power ratings in units of *apparent* power, units are Volt- Amps (or kiloVolt-Amps, kVA or MegaVolt-Amps, MVA).[2]

Utility Revenue Requirement for T&D Equipment

The T&D revenue requirement is the amount of revenue that utilities are required to receive from ratepayers to cover costs – hence the term *revenue requirement*.

Capital Plant Carrying Charges
Part of the revenue requirement – capital plant carrying charges (carrying charges) – covers the costs to own the plant and equipment (capital plant). The primary components of carrying charges include principal, dividend, and interest payments for the capital used to purchase and install the equipment. Insurance and taxes are also included.

Expenses
The revenue requirement also includes expenses; in this case, maintenance and operation expenses for T&D equipment. T&D expenses are usually quite small relative to annual capital plant carrying charges and are not addressed in this evaluation.

Annual Revenue Requirement
The *annual* amount of revenue needed to cover costs incurred *in a given year* is the *annual revenue requirement*. Typically, the annual revenue requirement is calculated by multiplying the installed cost for the T&D equipment by a utility-specific "fixed charge rate." The fixed charge rate reflects all elements of the carrying charges: annual payments for return of principal, interest, and dividend payments plus annual income tax, property tax, and insurance payments. In most cases, the fixed charge rate for T&D reflects an expected equipment life ranging from 20 to 40 years.

Typical fixed charge rate values range from 0.08 to 0.15. Lower values reflect the cost for public or cooperative utilities that: a) do not have stockholders who receive dividends, b) use public (debt) capital with relatively low interest rates, and c) do not pay taxes. Higher values reflect capital costs for investor-owned utilities: a) whose financing involves equity with dividend payments to stockholders and use of private sector debt, and b) that pay taxes.

A representative fixed charge rate of 0.11 is used in this report.

T&D Avoided Cost

In simplest terms, the T&D deferral benefit is the "avoided cost" – the cost *not* incurred by utility ratepayers if the T&D upgrade is not made. The avoided cost is equal to the revenue requirement.

In many cases, an upgrade deferral option is viable for one year or a few years. In those cases, the deferral benefit is the *annual* avoided cost (the annual revenue requirement) for the year(s) that the upgrade is deferred).

Often, when T&D facilities are upgraded, some portion of the equipment removed can be reused, especially transformers. Specifically, when a substation is upgraded and a transformer that can be re-used is removed from service to accommodate a larger one, the smaller transformer is returned to rotating stock. The financial (residual) value of the remaining useful life for re-used equipment is ignored in this report.

Single-Year Deferral Benefit

The *single-year* T&D upgrade deferral benefit is defined as the utility's *annual* revenue requirement for the upgraded T&D facility. Notably, if a T&D upgrade project *is* deferred, then the avoided payment is treated as if it is avoided forever.

Consider the working example: a T&D node will be upgraded from 12 MW to 16 MW, a capacity addition of 33% (4 MW). The 12 MW equipment is removed and replaced with equipment rated at 16 MW for a cost of $75 per kW installed. The entire project costs $7 5/kW * 16 MW * 1,000 kW/MW = $1,200,000. Using the representative fixed charge rate of 0.11, the *annual* charge to own the upgraded equipment is 0.11 * $1,200,000 = $132,000.

Multi-Year Deferral Benefit

In some cases, it may be technically viable and financially attractive to defer an upgrade for more than one year. In general, these are circumstances involving: a) slower than expected aggregate load growth, b) block load additions that do not materialize or that are delayed, or c) specific loads that are removed or that diminish.

In many cases, especially those where *demand* is growing, a conventional T&D upgrade becomes the most viable option after one or two years of deferral, primarily because the amount of storage needed to defer an upgrade increases significantly as load grows. Conversely, if *load* is growing, the deferral benefit *per kW of storage* diminishes rapidly from one year to the next because: a) the amount of storage needed increases each year, and b) the annual amount of T&D cost that can be deferred stays nearly constant. (See subsection of Section 2 entitled *Multi-Year Deferral Benefit for MES* for details).

Once it is determined that a T&D upgrade *is* the lowest cost option (e.g., in year 3) and once the upgrade occurs, then the storage could be left in place and used to provide other benefits. If the storage is transportable, it could be moved to another site to provide T&D deferral or other valuable benefits.

MES Transportability

Because the annual T&D deferral benefit per kW of MES tends to diminish from one year to the next, MES resources can be used for one or two years at one location, then moved to other locations to provide additional deferral and/or other benefits. In fact, transportable storage could be used more than once, in different locations, in the same year. For example, a location where peak demand occurs in the summer, followed by another location with a winter peak.

Even if a storage system is moved and re-used *once* during the life of the storage plant, the effect on the storage's cost-effectiveness may be dramatic. Of course, utilities must be able to disconnect, transport, and reconnect the storage with modest effort and cost to make such redeployments practical and cost-effective. (See the subsection of Section 2 entitled *MES Redeployment* for details.)

T&D Upgrade Marginal Cost

Readers should also be familiar with the *marginal* cost for T&D capacity – the cost per kW of capacity *added*. If nothing else, this value is notable because some related research results and policy-related evaluations – including those presented below – are expressed in terms of T&D marginal cost.

Consider the working example: the existing T&D equipment to be replaced has a rating of 12 MW. The capacity after the upgrade will be 16 MW and will cost $7 5/kW *nameplate*. The amount of capacity *added* is 4 MW so the *marginal* cost is $1,200,000 / 4,000 MW = $300/kW of T&D capacity *added*.

2. T&D Deferral Financial Benefit

Introduction

For this report, the T&D deferral benefit is defined as the *annual* cost that will *not* be incurred (cost that is avoided) if a given T&D project upgrade is deferred. For utilities, that amount is the annual revenue requirement: the amount of money that must be collected from utility ratepayers at large to cover the single-year cost.

Consider the working example introduced in Section 1: T&D equipment rated at 12 MW will be replaced with equipment rated at 16 MW, an addition of 33% more capacity, or 4 MW. The project will cost costs $75 per kW of T&D capacity *installed*. So the entire upgrade project costs $75/kW * 16,000 kW = $1,200,000.

Annual Deferral Benefit for MES

Using the representative fixed charge rate of 0.11, the annual charge (and revenue requirement) to own the upgraded equipment is 0.11 * $1,200,000 = $132,000. This is the money "in play" in a given year.

A helpful way to express the single-year deferral benefit attributable to storage – expressed in units of dollars per kW of storage capacity in the respective year – is calculated by dividing the annual T&D revenue requirement for the deferred upgrade by the kW of storage.

Consider the working example with load growth as shown previously in Figure 1. In year 0 (the year before overloading occurs), no storage is needed because demand is lower than the existing T&D capacity. Assuming that load will grow by 2% between year 0 and year 1, demand is expected to exceed the T&D capacity in year 1 by 87 kW.

Given that load projections are not precise and that most storage system vendors only accommodate specific capacity increments (e.g. 50, 100 or 200 kW), utility engineers would most likely oversize any modular resource used for deferral. In the working example, assume that 150kW of storage is specified to serve the 87 kW of expected overload in year 1. (That is about 72% more storage capacity than the expected overload.)

For the example, the first year deferral benefit (per kW of storage) is:

$132,000 revenue requirement ÷ 150 kW of storage = $880/kW of storage

Multi-Year Deferral Benefit for MES

Continuing with the example shown in Figure 1, in year 1 demand is expected to exceed the T&D capacity in year 1 by 87 kW. Assuming that the load grows another 2% between years 1 and 2, the projected load will exceed T&D capacity by about 329 kW in year 2.

To serve that 329 kW expected overload in year 2, assume that engineers specify a storage system rated at 500kW to defer the upgrade in the second year (52% more than the expected overload, and 230% more than the 150kW used for deferral in the previous year).

If ignoring the time value of money, the second year deferral benefit (per kW of storage) is:

$132,000 revenue requirement ÷ 500 kW of storage = $264/kW of storage

Compare that single year deferral benefit in year 2 ($264 per kW of storage) to the single year deferral benefit in year 1 ($880 per kW of storage).

Results from the calculations for years 1 and 2 are summarized below in Table 1.

Table 1. Summary Results for Two-Year Deferral Case Example

	Power Rating	Portion of Existing T&D Capacity (%)	$/kW
Existing	12.0		
Upgraded	16.0		75.0
Added	4.0	+33.3%	300.0
Year 0	11.85	98.8%	
Year 1	12.09	100.7%	
Year 2	12.33	102.7%	
Year 1	87	+0.7%	1,517*
Year 2	329	+2.7%	402*
Year 1	150	1.3%	880**
Year 2	500	4.2%	264**

Upgrade Total Cost: 16.0 MW * $75/kW = $1,200,000. Upgrade Annual Cost: $1,200,000 * 0.110 = $132,000. * Upgrade annual cost ÷ Overload. Units: $/kW-year. ** Upgrade annual cost ÷ MES Capacity. Units: $/kW-year.

Diminishing Benefit for Multi-Year Deferrals

As illustrated in the example above, the deferral benefit per kW of storage can diminish dramatically from one year to the next. So, in almost all cases, at some point in time (probably after one or two years of deferral) the T&D upgrade will be more cost-effective than adding MES. That topic is addressed in more detail later in this section of the report.

There are two fundamental drivers of this phenomenon. First, from one year to the next load grows, so more storage is needed each year. Second, while the amount of load exceeding the T&D capacity may increase considerably each year, normally the annual financial benefit for T&D deferral (i.e., the avoided revenue requirement) stays almost constant (without regard to the time value of money).

The effect is even more dramatic if a five year horizon is considered, as shown in Figure 2.

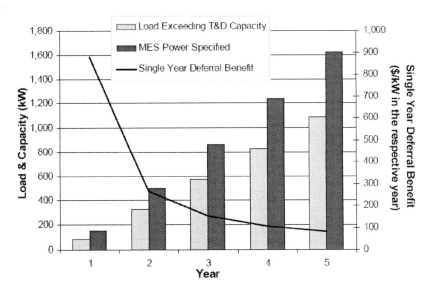

Figure 2. Diminishing Deferral Benefit per kW of MES Over Five Years.

MES Redeployment

If the MES is transportable and can be redeployed, the possible financial implications are compelling. Consider Figure 3 which is based on the working example above.

The value proposition represented in Figure 3 is as follows: storage is used at five different locations for one year of T&D upgrade deferral, in alternating years, over a ten year span. In the other five years – when the storage is not used for T&D deferral – it provides a benefit related to improving local power quality (PQ) and/or electric service reliability (reliability).

Each of the five upgrade projects deferred with the transportable MES system will involve deferral of a project to replace equipment rated at 12 MW with equipment rated at 16 MW. Each T&D upgrade will cost $75 ($Year 1) per kW of T&D capacity installed, or $1.2 million. The single year deferral benefit at each of the five locations is $1.2 million * 0.11 = $132,000.

The projected overload at each location (in the year of the deferral) is assumed to be 2% of the existing capacity, or 12 MW * 2% = 240 kW. The amount of storage required for a single year upgrade deferral is assumed to be 50% more than the expected overload (1.5 * 240 kW = 360 kW). So, the single year deferral benefit for storage at each of the five locations is $132,000 ÷ 360 kW = $367/kW of storage ($Year 1).

In the five years when storage is not used for deferral, it is assumed to be located so that it provides a benefit of $75/kW[3] of storage ($Year 1) for improving local power quality (PQ) and/or electric service reliability (reliability).

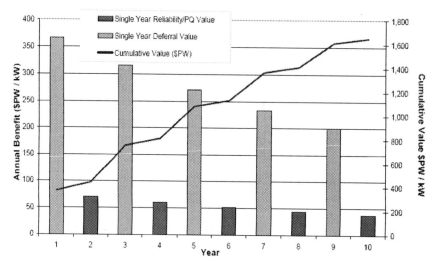

Figure 3. Value Proposition for Transportable Storage.

As shown in Figure 3 (on the right-side Y axis), the present value of the annual benefit is nearly $ 1,700/kW of storage (assuming 2.5% annual inflation and a 10% discount rate). So, if storage can be purchased, owned, and operated for less than $ 1,700/kW, for ten years, then it would be a financially attractive option. That value would provide a helpful target for lifecycle costs for MES (in this case, with a ten-year life).

T&D Equipment Residual Value

Often, existing T&D equipment is removed to accommodate an upgrade; though, in many cases, the equipment removed could still provide service elsewhere in the utility's system. In those cases, the equipment removed is placed into the utility's rotating stock.

From the example described above, assume that the upgrade equipment installed is a transformer rated at 16 MW and that it replaces an existing transformer rated at 12 MW. The 12 MW transformer: a) has 15 years of useful life remaining, b) originally had 40 years of life, and c) originally cost $30/kW ($360,000 total).

For this report, residual value is ignored. In effect, that treats residual value as if: a) it is a sunk cost that cannot be deferred or avoided, and thus b) it does affect the benefit for deferral.

T&D Annual Cost Among U.S. Electric Utilities

As one might expect, the average value for T&D deferral benefit varies widely among utilities, locations, and from year to year. A good source for data about the average cost for various utilities is "FERC Form 1". FERC is the Federal Energy Regulatory Commission. FERC Form 1 contains detailed information about investor-owned utilities' investments in capital equipment, including T&D equipment.

In simplest terms, what could be called the average marginal cost for T&D capacity is calculated by dividing annual expenditure data from FERC Form 1 by the sum of the nameplate ratings of all load carrying capacity added.

FERC Form 1 Data: Oak Ridge National Laboratories Report

One of the best summaries of FERC Form 1 data of which the authors are aware was produced by Oak Ridge National Laboratories (ORNL).[1] The report evaluated average marginal cost for 105 utilities in the U.S., with a breakout of utilities in the combined Pennsylvania, New Jersey, and Maryland region (the area of the country served by PJM Interconnection (PJM), the regional transmission organization (RTO)).

Table 23 in the Oak Ridge report (shown on the following page as Table 2) provides the breakout value of distribution equipment costs for major utilities in 1998 as reported in FERC Form 1. The table shows data for accounts 360 through 368 which include utility distribution equipment. [1]

Table 2. End-of-Year Balance of Distribution Equipment for 105 Utilities and for 11 PJM Utilities

FERC Form1 Account	Marginal Cost ($/MVA) National 1989 to 1998	Marginal Cost ($/MVA) PJM 1989 to 1998	Average Cost ($/MVA) National 1998	Average Cost ($/MVA) PJM 1998
Dist Land (360)	2,639	5,653	1,501	2,978
Dist Structures (361)	2,481	5,538	1,219	3,408
Dist Station Equip (362)	32,869	57,248	16,925	25,820
Dist Battery Storage (363)	2	0	0	0
Dist Poles & Towers (364)	50,390	50,746	22,403	24,457
Dist Overhead Conduct (365)	52,059	63,363	22,246	28,366
Dist Undgr Conduit (366)	13,815	23,739	6,428	12,376
Dist Undgr Conduct (367)	44,226	65,121	18,043	26,885
Dist Transformers (368)	40,787	39,757	23,656	24,715
Dist Services (369)	26,553	34,494	11,888	16,433
Dist Meters (370)	13,625	14,045	7,655	8,989
Dist Installations (371)	2,854	4,858	1,133	1,327
Dist Leased Property (372)	-131	1	42	6
Dist Street Lights (373)	8,034	10,175	4,438	4,610
Dist Total	290,203	374,737	137,576	180,369
Trans Total	80,650	64,876	52,229	48,681
Total Dist and Transmission	370,853	439,613	189,805	229,050

Source: The data is from 105 utilities selected from the intersection of utilities for these accounts in both 1989 and 1998 included in the POWERdat database (Resource Data International, Inc.). This data was originally from data collected in FERC Form 1.

The values of interest for this analysis are those in the columns labeled "Marginal Cost." This is the cost for equipment *added* (on the margin), whereas the values in the columns labeled "Average Cost" reflect the cost for *all* equipment in place, including equipment that is "fully depreciated" and which therefore has an annual cost of zero.

Continuing from the ORNL report:

"An approximation of the marginal cost for new equipment was determined by taking the difference in the account EOYB's [end-of-year balances] for 1998 and 1989 divided by the difference in the distribution capacity for the same period. (The marginal rates tend to vary greatly from year to year, so a ten-year time span was used to get a reasonable average value.) The distribution capacity was taken to equal the Distribution Line Transformer capacity.

Over this period, the marginal cost of *distribution* capacity was $290,203/MVA ($290/kVA) nationally and, $374,737/MVA ($375/kVA) for the 11 PJM utilities. [Notably] the average *marginal* cost of distribution capacity over this period was more than double the average *embedded* cost.

The marginal cost for *transmission* for the same period was $80,650/MVA ($80.6/kVA) for the 105 nationwide utilities and $64,876/MVA ($64.9/kVA) for the 11 PJM utilities."[1]

As context, recall the working example (Section 1) with a cost of $75 per kW installed for an upgrade that increases T&D capacity from 12 MW to 16MW, an increase of 33% or 4MW. In that example, the marginal cost for capacity is $1.2 Million/4,000 kW = $300/kW. That value ($300/kW marginal cost) is comparable to the values developed by ORNL for distribution capacity ranging from $290/kVA nationwide and $375/kVA for the PJM area (without regard to power factor).

Rate Assistance Program Report

Another resource for information about T&D marginal costs is the Rate Assistance Program (RAP). (Details are available at raponline.org.) Consider these findings and observations from RAP's "review of the cost of expansion of the distribution system for all utilities filing a FERC Form 1 for the period 1994-1999."[2]

- "The average marginal costs for transformers and substations ranged from virtually zero to over $3,500 per kW." After excluding areas where load did not grow or where demand contracted, "the average was $136 per kW, with a standard deviation of over $356 per kW."
- With regard to circuits (wires), marginal cost "ranged from virtually zero to as high as $19,483 per kW, for an average of $872 per kW, with a standard deviation of over $2,800 per kW."

Note that most T&D projects – certainly the more expensive ones – probably include some combination of new transformers and wires capacity.

Utility T&D Deferral Cost Variability

The Range of T&D Deferral Benefits

The previous section of this report provided a good indication of the *average* costs for T&D capacity added. Specifically, that value reflects the *average* cost for adding T&D capacity, over a period of several years, at all locations requiring an upgrade in a given year.

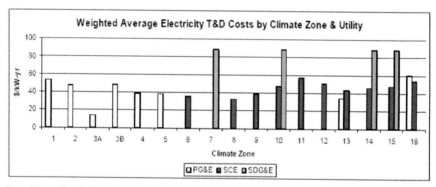

Note: Climate Zone 3A includes San Francisco, East Bay, and Peninsula sub-areas, while 3B includes portions of Central Coast, Mission, and North Bay.
(Source: California Public Utilities Commission[3]).
(Note: values in Figure 4 are expressed in units of $/kW per year of T&D equipment, whereas average marginal cost values in the previous subsection are expressed in units of $/kW T&D capacity installed. To convert $/kW installed to $/kW-year, use the fixed charge rate. If using the representative fixed charge rate of 0.11 and an average marginal installed cost for T&D equipment of $3 00/kW, the annual value is 0.11 * $300/kW = $33/kW. As context: based on the values in Figure 4, T&D marginal cost in California is about twice the average marginal cost nationwide).

Figure 4. Weighted Average Annual T&D Avoided Cost for Large Investor-Owned Utilities in California.

Depending primarily on geographic conditions and variation in the area evaluated, the marginal cost for specific projects (marginal cost) may vary by a factor of ten or more. This variation is driven by criteria including the population density of the area, terrain, geology, weather, and the type and amount of T&D equipment involved (especially projects with a large footprint or that involves underground circuits).

T&D marginal costs in California vary by a factor of seven for the three large investor-owned utilities (PG&E, SCE, and SDG&E). The percentage of those costs that are related to peak demand during the summer can vary by up to 103%. The analysis "indicates that climate zone is the dominant [indicator]"of T&D marginal cost. [3]

Figure 4, below, displays the weighted average annual T&D avoided cost, by climate zone, for those three major utilities in California.

Similar results (expressed in units of $/kW installed) were published in The Energy Journal. [4] An overview of results from that study provides another good indication of the range of benefits for T&D deferral. Results reflect variations observed for Pacific Gas and Electric (PG&E) and for Public Service of Indiana (PSI). Values reflect the costs associated with adding the equipment needed to accommodate demand growth.

Key Results from The Energy Journal study:

PG&E

- T&D deferral benefits vary widely among PG&E's distribution planning areas; 19% of the areas have zero T&D deferral value.
- The average and maximum T&D deferral benefit values are $230/kW and $1,173/kW, respectively.

PSI

- T&D deferral benefits vary widely among PSI's distribution planning areas; 73% of the areas have zero T&D deferral value.
- The average and maximum T&D deferral benefit values are $64/kW and $1,040/kW, respectively.

Note that the average deferral benefit between 1994 and 1999 varies modestly for both utilities, though the deferral benefit varies dramatically among planning areas. Note also that the values above are expressed in $1999.

EPRI/PG&E Distributed Utility Penetration Study (DUPS)

The range of T&D deferral benefits was examined in the Distributed Utility Penetration Study (DUPS), a detailed evaluation of the Pacific Gas and Electric Company (PG&E) distribution system.[5] The evaluation was jointly funded by PG&E and the Electric Power Research Institute (EPRI). From the report:

"The objective is to measure the monetary savings that can be achieved by PG&E if all cost-effective applications of distributed utility (DU) devices, direct load control and demand-side management programs are adopted. DU devices include batteries, gensets, fuel cells, photovoltaics, direct load control, and demand-side management programs."

Based on results from the EPRI study: in 50% of locations in PG&E's service area that require distribution upgrades in any given year, the marginal cost is at least $381/kW *added*. For the most expensive locations requiring upgrades (90th percentile and above), the cost exceeds about $600/kW *added*. (All values adjusted to $2004 basis.) The distribution of marginal costs – for distribution capacity added – is shown in Figure 5.

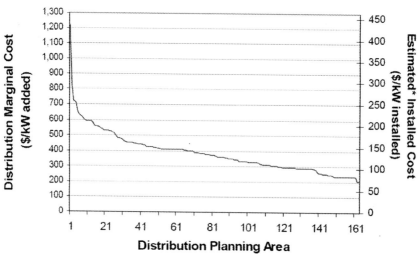

* T&D Upgrade Factor.
Sources: PG&E and EPRI or = 0.33.

Figure 5. PG&E T&D Distribution Marginal Costs.

Note that the above values reflect distribution capacity marginal costs for locations where there is an existing or emerging need for an upgrade. Locations that do not require upgrades (i.e., with marginal cost of 0) are not included.

Importantly, those marginal cost values are for capacity *added*. A general indication of the relationship between marginal cost ($/kW of T&D capacity added) and installed cost ($/kW nameplate of T&D equipment installed) is provided by applying the generic upgrade factor from the working example.

Consider the mean marginal cost of $381/kW for distribution capacity from the PG&E data. The corresponding installed cost is $381/kW * 0.33 = $125.7/kW nameplate, for T&D equipment installed. Further, 10% of all upgrades required have a marginal cost above about $600/kW installed or $600/kW * 0.33 = $ 198/kW nameplate. (For comparison, the working example is based on T&D capacity installed cost of $75/kW nameplate.)

Finally, though the magnitude of PG&E's distribution marginal costs may be higher than those for most other utilities, given the wide diversity of geographic conditions and load types in PG&E's service territory, the spread and perhaps even the variability of PG&E's distribution marginal cost may provide a helpful indication of the variability of that cost elsewhere.

3. DEPLOYMENT SITUATIONS FOR T&D DEFERRAL APPLICATIONS *INTRODUCTION*

This section describes circumstances for which T&D deferral, using modular distributed energy resources (DERs) including modular electricity storage (MES), yields an attractive financial benefit.

At the highest level, those circumstances involve a node or location within a T&D system where peak demand does or will soon exceed the T&D equipment's rating – a "hot spot." Presumably, locations (hot spots) for which DERs *could* be an attractive alternative are identified during the normal T&D capacity planning cycle. The next step is to identify hot spots for which DERs may be the lowest-cost option to serve peak load *on the margin* during the next year.

In broad terms, T&D deferral using modular resources is attractive if: 1) it reduces total cost-ofservice to utility ratepayers, 2) capital *not* used – because the T&D investment is deferred – is used for other important T&D projects, 3) doing so increases utility asset utilization, or 4) doing so reduces financial risk.

Decision Criteria Summary

The following criteria indicate whether a specific hot spot is well-suited for T&D deferral using modular resources (listed in no particular order):

- High T&D upgrade cost (i.e., on a $/kVA of capacity-added basis) – including direct costs and "soft" costs such as utility reputation and customer goodwill.
- High peak-to-average demand ratio; such locations are sometimes said to have a "peaky" diurnal load profile.
- Modest projected overload – peak demand is expected to exceed the T&D system's rating by a modest amount.
- Slow peak demand growth (rate).
- Uncertainty about the timing and/or likelihood of *block* load additions.
- T&D construction delays or construction resource constraints may be a challenge.
- Budget optimization – a T&D upgrade project competes with other important projects for capital.
- Benefits aggregation – the same distributed resource provides additional benefits (revenue or avoided cost) that can be aggregated into an attractive total value proposition, such as:

    o on-peak energy
    o electric supply capacity
    o value enhancement for electricity from renewable energy resources o reduced transmission congestion and energy losses
    o electricity end-user energy/demand bill reduction
    o electric supply reserve capacity
    o improved local power quality and/or reliability

MES is especially well-suited to these hot spot locations if:

- Air emissions regulations, noise regulations, fuel storage or other safety-related challenges restrict use of distributed *generation*.
- The price differential is large between times when storage is charged and when it is discharged.
- Used to enable end-user participation in demand management or curtailable load programs.

High T&D Upgrade Cost

Probably the most obvious criterion affecting the merits of modular resources is the cost of the T&D project to be deferred. In general, T&D projects whose unit cost (per kW or per kVA) is high tend to be superior candidates for deferral using modular resources.

High Peak-to-Average Demand Ratio

Modular resources are especially well-suited to locations where peak loading has a relatively short duration. Those are locations with a relatively high peak to average load ratio. Such locations are sometimes said to have "peaky" demand (profiles).

The plots in Figure 6 illustrate this characteristic. Profile #1 is the narrowest and "peakiest", while the profile with the broadest, least peaky maximum is profile #3.

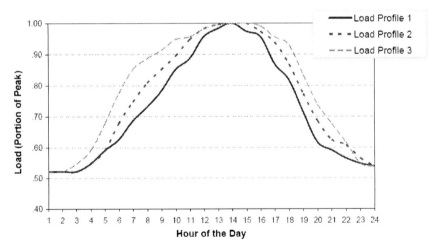

Figure 6. Load Profiles Illustrating "Peakiness".

For storage, the primary reason load "peakiness" is important is that "peakier" loads require shorter discharge durations, and thus less energy storage equipment per kW of storage system power. For distributed generation, "peakier" demand means less run-time. For demand management, "peakier" loads mean that less load is "managed" for fewer hours per year.

Modest Projected Overload

Generally, DER and MES are more attractive for situations involving relatively small projected overload (demand exceeding rated capacity) because a relatively modest amount of storage is required.

Projected overload (kW or MW) is a function of: a) the expected peak demand growth during the next year, and b) T&D capacity slack. T&D capacity slack is the difference between: a) the maximum demand during the previous year, and b) the load carrying capacity (rating) of the T&D equipment serving that load.

Capacity slack varies among hot spots. This concept is illustrated graphically in Figure 7. It shows a range of capacity slack values for T&D nodes whose peak demand is approaching the T&D equipment's load carrying capacity.[11]

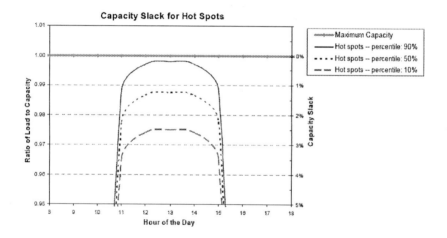

Figure 7. Variability of T&D Loading Relative to T&D Rating, Among Hot Spots.

Consider an example: two very similar distribution system hot spots with a rated capacity of 12,000 kW and 3.0%/year annual load growth. In both locations, the upgrade will involve adding 4,000 kW.

Assume that both upgrades will cost $1,200,000 (that is, $1.2 Million ÷ 12,000 kW = $100/kW of T&D capacity *installed* or $1.2 Million ÷ 4,000 kW = $300/kW of T&D capacity *added*). The *annual* cost (revenue requirement) for both upgrades is $132,000 per year (assuming a fixed charge rate of 0.11 to calculate the annual carrying cost for the initial investment: 0.11 * $1.2 Million).

At the first hot spot, peak loading in the previous year was 98% of the equipment's rated capacity, and at the second hot spot, the previous year's peak load was almost the same as the T&D equipment's rated capacity.

At the first location peak demand growth of 3.0% yields a projected overload of about 1.0% in the next year, or 12,000 kW of existing capacity * 1.0% = 120 kW. The benefit for using exactly 120 kW of modular resources to defer the T&D upgrade is about $132,000 ÷ 120 kW = $1,100/kW, for one year of deferral.

At the second location, peak demand growth of 3.0% causes a projected overload of about 3%, or 12,000 kW of existing capacity * 3.0% = 360 kW. The benefit for using 360 kW of modular resources to defer the T&D upgrade is $132,000 ÷ 360 kW = $367/kWstorage. This phenomenon is illustrated graphically in Figure 8.

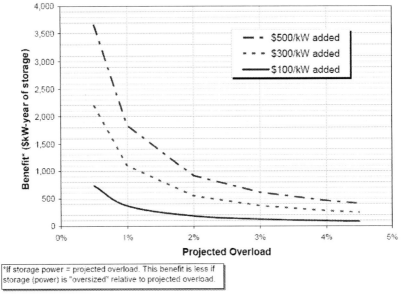

Figure 8. One-Year Deferral Benefit from MES as Function of Projected Overload, 0.11 Fixed Charge Rate.

Slow Load Growth

In general, it is more beneficial to defer upgrades using modular resources at hot spots where peak demand is growing slowly than it is to defer upgrades of T&D equipment serving demand that is growing rapidly (specifically, *inherent* demand growth, not including block load additions such as new commercial facilities and residential development). There are at least three reasons for this:

- If planners are somewhat-to-very certain that demand growth will be slow, then a low demand growth rate usually indicates the need for a relatively small amount of modular resources to defer an upgrade, in a given year.
- Because relatively small amounts of modular resources are needed – in a given year – modular resources may be economically viable for more years of deferral, if demand growth is low.
- If planners are somewhat-to-very certain that demand growth will be slow, then using modular resources may have relatively low risk (compared to using modular resources at hot spots where demand growth is high) because there is less chance of an expensive overload of T&D equipment if the modular resources fail or are undersized.

Uncertainty About Block Load Additions

MES and other DERs may be attractive alternatives when there is uncertainty about the magnitude and timing of block load additions that would cause an overload. Block load additions are usually related to commercial or residential development or to expansion of existing industrial facilities.

If there is a low probability that a block load will materialize before the next peak demand season, then modular resources may be an attractive option. If there is a chance that the block load will never materialize, then modular resources may be even more attractive. Modular resources used to address uncertainties in block load additions may be especially attractive if T&D construction delays are also likely.

Uncertainty About T&D Construction Delays

For a planned T&D upgrade, construction delays – that lead to T&D equipment overloading – are possible. Delays may occur for several reasons, including budget or other resource constraints, or institutional reasons such as permitting.

In addition to capital budget shortfalls, other resource shortages that could lead to construction delays include: a) land, b) construction equipment, c) T&D equipment, d) engineering, e) construction, and f) other staff costs.

Institutional hurdles that could cause uncertainty, and possibly construction delays, may include: a) construction, electrical, and fire permits, b) environmental impact reports, and c) land use/zoning.

If there is a significant chance that construction will be delayed – for any single reason or multiple reasons – then modular resources could be an attractive alternative to a conventional T&D upgrade.

Distribution Budget Optimization

One way that modular resources can reduce utility costs overall is that they provide utilities with a rich spectrum of possible responses to a wide range of location-specific T&D capacity needs on the margin. Given a high degree of modularity and the various modular electricity technology types, the number of possible combinations is large and growing.

In a given year, there may not be enough capital and/or labor resources available to undertake all upgrade projects identified by distribution engineers. In those situations, modular resources could be used to increase available options. That could, in turn, be used to optimize the "portfolio" of responses, including: a) capital investments in new equipment, and b) expenses such as equipment rental or leasing and fuel.

Consider an example: distribution engineers identify seven locations requiring upgrades; however, there is only enough capital to pay for five of the seven, and there are only enough labor resources (person-hours) available for six projects.

The key driver at location five is a housing development that will add 5% to the peak demand, 3% above the T&D equipment's rating. However, it is likely that less than half of the new load will be connected before the end of the peak-demand season. So, storage rated at 1% of the expected peak demand is used to defer the upgrade at location five, which frees up enough capital and labor to proceed with the upgrade at the sixth location.

At location seven, utility account representatives contract with a commercial customer under terms of a curtailable load program, so the needed upgrade at that location may be deferred.

Benefits Aggregation for Attractive Value Propositions

In many circumstances for which MES is *technically* viable, it may not be cost-effective unless more than one benefit can be aggregated to yield a relatively high total benefit. So, locations/circumstances for which MES provides two or more benefits may be superior prospects, given total benefits. Other benefits are described in a report published by Sandia National Laboratories entitled *Energy Storage Benefits and Market Analysis Handbook.* [6]

4. STORAGE CHARACTERISTICS FOR T&D DEFERRAL *INTRODUCTION*

The modular electricity storage (MES) characteristics needed for T&D deferral are listed below. A discussion of each characteristic follows.

There are at least four required characteristics:

- Reliability
- Power Quality
- Ramp Rate
- Charge Rate

Other notable characteristics include:

- Roundtrip Efficiency
- Maintenance Needs and Cost
- Emergency or Short Duration Power Capability
- Lifetime Charge/Discharge Cycles
- Self-discharge Rate
- Energy and Power Density
- Plant Footprint and Volume
- Modularity
- Transportability
- Easily Deployable ("Plug and Play")

Required Characteristics

Reliability

Reliability is the most important MES performance criterion for T&D deferral. Before power engineers will use MES for T&D deferral, they must be assured that the MES equipment will perform as expected, when needed.

Given MES with the necessary quality and appropriate design, one approach to high reliability is *modularity*. By using several smaller units operated as an aggregated system, reliability is enhanced because it is very unlikely that all, or even most, capacity will not be functioning correctly at any given time.

Further, MES for T&D deferral is used infrequently. That reduces the chance that the MES will not perform adequately because: a) there are fewer opportunities for equipment failure, b) a lower frequency of MES use reduces "wear and tear" on the equipment, and c) for some types of storage, limiting the depth of discharge also reduces wear and tear. Conversely, using the same MES for T&D deferral and for one or more other benefits may reduce MES reliability.

Power Quality

The quality and characteristics of the power delivered from MES is quite important. At worst, the MES should "do no harm" to the quality of power delivered by the utility. Under some conditions, MES may even be expected to improve power quality (PQ).

The MES subsystem that has the greatest effect on PQ is the power conditioning unit (PCU). The most important element of the PCU is an inverter that converts DC power from storage to AC power for loads.

Wave Form

Utility power is produced and delivered in alternating current (AC) form. In AC power systems, the magnitude of the power being delivered oscillates between a maximum value with a positive polarity and the corresponding maximum value with negative polarity. In the United States, that cycle occurs at a rate of 60 times per second (a.k.a. 60 Hertz or 60 Hz).

The power oscillations occur continuously, and they occur with a smooth and predictable pattern, depicted graphically by the "sine wave," as shown in Figure 9.

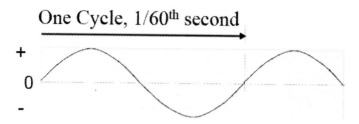

Figure 9. 60 Hz Sine Wave.

Lower cost inverters whose output has the form of a square wave (shown graphically in Figure 10) or modified square wave, are generally not suitable for situations requiring utility-grade power.

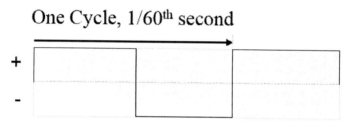

Figure 10. 60 Hz Square Wave. Voltage Stability.

MES should provide output whose voltage remains stable under the range of expected operating conditions.

High Power Factor

MES systems used for utility-related applications should deliver real power at a power factor that is as close to unity (1.0) as possible. This obviates the need for utilities to install equipment for power factor correction within the T&D system. In fact, utilities are likely to establish minimum standards for power factor for any equipment that interacts electrically with the T&D system.

Frequency Stability

Typically, the frequency of utility power varies only modestly (much less than 1%). So, MES used for T&D deferral should have very stable output frequency under the full range of normal operating conditions.

Harmonics

Harmonic currents in distribution equipment can pose a significant challenge. Harmonic currents are components of a periodic wave whose frequency is an integral multiple of the fundamental frequency. In this, case the fundamental frequency is the utility power line frequency of 60 Hz. So, for example, harmonic currents might exist with frequencies of 3 x 60 Hz (180 Hz) or 7 x 60 Hz (420 Hz). Total harmonic distortion (THD) is the contribution of all the individual harmonic currents to the fundamental frequency.

A large portion of harmonic currents are injected into the grid by modern electronic equipment with "switch mode" power supplies that convert AC power at line voltage to low voltage DC power used by electronic circuits, using solid state power switching devices. The most common of those "non-linear" loads are consumer electronics, computers, and uninterruptible power supplies (UPSs). Other important sources of harmonics are variable speed (motor) drives and electronic lighting ballasts.

MES should operate with total harmonic distortion (THD) of 5% or less.[12]

Ramp Rate

Ramp rate is the rate at which power output can change. For storage, the rate at which charging can vary may also be important. Generally, storage ramp rates are high (i.e. output can change quite rapidly.)

Under almost all conditions, the output from most T&D equipment (i.e., wires and transformers) changes nearly instantaneously. By contrast, power output from most types of generation changes slowly. This is especially true for central power plants with rotating machinery (i.e., turbines and generators) that have a large mass.

Depending on the mode of operation, the ramp rate of MES used for T&D deferral may or may not be important. In some cases, MES may be operated at full output whenever maximum demand is expected. In those cases, the ramp rate is irrelevant. In other cases, the MES's power output "follows" demand as the demand changes. In those circumstances, MES output must change rapidly, to accommodate demand changes, or T&D equipment may become overloaded, leading to outages or to T&D equipment damage. Ideally, MES will respond within a few cycles.

Charge Rate

This criterion is important because, often, MES must be recharged so it can serve load during the next discharge period. If storage cannot recharge quickly enough, then it will not have enough energy to provide the necessary service.

In most cases, storage charges at a rate that is similar to the rate at which it discharges. In some cases, storage may charge more rapidly or more slowly, depending on: a) the capacity of the power conditioning equipment, and b) the condition, chemistry, and/or physics of the energy storage medium.

Other Notable Characteristics

The storage characteristics described briefly below may or may not be important depending on the value proposition (combination of benefits) for which storage is used.

Roundtrip Efficiency

Round trip efficiency – or just efficiency – is the amount of energy that is discharged for each unit of energy used for charging. Typically storage efficiency is between 60% and 90%, depending primarily on the type of storage medium used. Notably, for T&D deferral, efficiency is relatively unimportant because there are relatively few hours during the year when the storage would be used. More generally; this criterion is important if storage must be discharged for more than a few hundred hours per year.

Operation and Maintenance Needs and Cost

Operation cost is primarily related to staffing needed to operate the system. For smaller systems, operation cost is quite low, unless staff is needed during storage operation. Larger systems may indeed require staff to monitor and manage operation. In most cases, the operation cost per kWh of energy delivered (from storage) is low.

Maintenance cost is the cost incurred to repair and maintain a storage system. There are two categories of maintenance cost: fixed and variable.

Fixed maintenance includes all annually recurring costs incurred irrespective of the amount of storage operation. Consider, for example, the cost for quarterly or annual storage system inspection and diagnostics. Another example is routine maintenance of the storage system's enclosure (if any) or routine maintenance of supporting systems such as cooling needed for some storage types when used in regions with high temperatures.

Variable maintenance cost is proportional to the amount of storage operation. Variable maintenance cost is dominated by contributions to a "sinking fund" that is used to replace various parts of the storage that degrade with use, especially electrolyte and cells. Depending on the circumstances, cost for staff needed to perform variable maintenance may also be significant.

Emergency or Short Duration Power Capability

In some circumstances, it may be helpful if MES can provide "extra" power during situations involving an urgent, short duration, and infrequent need for power beyond the normal output of the MES. One example is the sodium/sulfur (Na/S) battery type. A Na/S battery, like many battery technologies, is capable of producing two times its rated (normal) output for relatively short durations.[7] Depending on circumstances, such a capability could be valuable.

Lifetime Charge-Discharge Cycles

To one extent or another, most energy storage media degrade with use (i.e., during each charge- discharge cycle). The rate of degradation depends on the type of storage technology used, operating conditions, and operating profiles.

For some types of electrochemical batteries, the extent to which the system is emptied (depth of discharge) may also affect the storage media's useful life. Discharging a small portion of stored energy is a "shallow" discharge, and discharging most or all of the stored energy is a "deep" discharge. Typically, a shallow discharge is less damaging to the storage medium than a deep discharge. The frequency of discharge can also have a significant impact on battery life.

Note that many battery vendors can produce storage media with extra service life (relative to the baseline product) to accommodate additional charge-discharge cycles and/or deeper discharges. Of course, there is usually corresponding incremental cost for the superior performance.

To the extent that the storage medium degrades and must be replaced during the expected useful life of the MES, the cost for that replacement must be added to the variable operating cost of the storage system.

Standby Losses

To one extent or another, most types of storage media have what may be generically called standby losses (energy losses that occur when the storage is charged but not being used). Specifically, electrochemical storage may tend to "self-discharge" after the storage is charged and while the storage is not in use (i.e., while the storage is not being discharged). Depending on the amount of time that passes between storage discharging, temperature, and other factors, this characteristic could have a non-trivial effect on the overall cost-effectiveness of storage for specific applications.

Energy and Power Density

Energy density is the amount of energy that can be stored in a MES device with a given volume. Similarly, power density is the amount of power that can be delivered from an MES with a given volume. These criteria are important in situations where space is valuable or limited.

Plant Footprint and Volume

Depending on the availability of floor space or property area, the MES footprint and space (volume) requirements may be important. To some extent, these criteria are driven by energy density and power density.

Modularity

One attractive feature of distributed energy resources in general, and modular electricity storage specifically, is the flexibility that modular "building blocks" provide. Modularity allows for more optimal levels and types of capacity, especially on the margin. With modular resources, utilities can increase and decrease capacity, on the margin, when and where needed, in response to changing conditions. Modularity – especially when combined with rapid deployment – provides utilities with attractive means to manage risk associated with utility T&D investments because capacity can be added (or removed) in relatively small increments, and utilities can therefore make more "nimble" responses to uncertainty.

Transportability

As discussed in detail in Section 2, for many T&D deferral projects, benefits are temporary, lasting for one or two years after which T&D upgrade investment may be the most attractive option. So, MES transportability greatly enhances the prospects that lifecycle benefits will exceed lifecycle cost.

Easily Deployable ("Plug and Play")

If MES is to be used at various locations, in addition to being transportable, the equipment should also be readily deployable once it arrives at a given location. To some extent, that requires engineering or regulatory protocols, procedures, practices, and rules which provide permission to utilities to connect MES to the T&D system and that specify appropriate way(s) for doing so.

5. STORAGE SIZING FOR T&D DEFERRAL

The following discussion is a summary of power-related and energy-related considerations for sizing energy storage used for T&D deferral. For more details, readers are encouraged to refer to a report published by Sandia National Laboratories entitled *Estimating Electricity Storage Power Rating and Discharge Duration for Utility Transmission and Distribution Deferral.* [13]

Storage Power Output Requirements

As illustrated in the working example: to defer a T&D upgrade for one year, the energy storage power output must equal the overload that engineers forecast (based on estimated load growth and the previous year's peak demand), plus a prudent allowance for uncertainty (primarily, uncertainty about load growth that may exceed expectations).

For most circuits, the highest loads occur on just a few days per year, for just a few hours per year. In some cases, the highest annual load occurs on one specific day whose peak is somewhat higher than any other day.

Storage Discharge Duration Requirements

Discharge duration is the amount of time that a storage plant must discharge at full power. Some utilities have predefined "design load profiles."

If using an actual time-varying demand profile (data), then the profile(s) for the day(s) with the highest and "widest" peak are used to develop a design demand profile(s). Demand profiles for several years may be evaluated to identify the profile with widest peak demand.

The concept is shown graphically in Figure 11, Figure 12, and Figure 13, below. These figures show the load profile on the peak demand day in year 0, along with plots that reflect expected load in years 1 and 2.

Figure 11 includes three plots. The first is the load profile for the peak day in year 0. The next plot shows the same profile with one year of load growth added. The final plot shows the same profile with another year of load growth added. Figure 12 provides a more detailed view of those plots. They emphasize the portion of load that is on the margin of the T&D equipment's load carrying capacity.

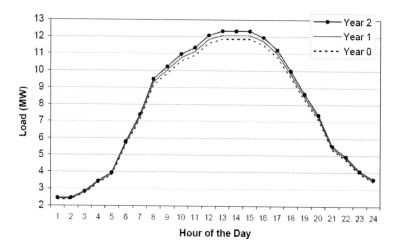

Figure 11. Load Shape on Peak Demand Day, Years 0 to 2.

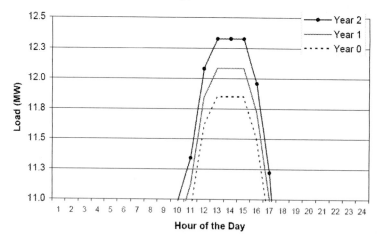

Figure 12. Peak of Load Shape on Peak Demand Day, Years 1 to 2.

Figure 13, below, shows the energy and power required for years one and two. In year 1 load exceeds the T&D limit from about 12:30pm to about 3:30pm. The most conservative discharge duration is about 3 hours.

A more precise calculation of the discharge duration can be made by estimating the total energy needed as a function of the area: a) under the demand curve, and b) above the T&D limit, as shown in Figure 13.

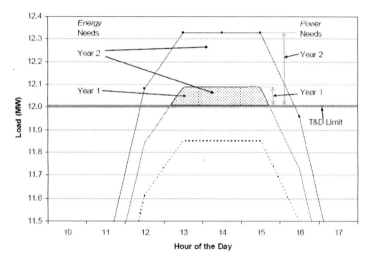

Figure 13. Storage Energy and Power Requirements.

Diesel Generation for Storage Backup

There is always a possibility that any modular resource may fail to operate as needed. Another challenge for all modular resources is that they may be undersized with respect to power. Storage poses yet another challenge because it is also energy-limited – once all stored energy is discharged, the storage is not useful as a power source until it is recharged.

One way to reduce the related risk is to retain the option to deploy diesel generation if load cannot be served adequately by the primary modular resource(s) used. Depending on proximity to dealers, diesel generation can be readily rented and quickly deployed – sometimes within a few hours and often within a day.

Though annual run hours for gensets (diesel engine/generator systems) used this way would be quite limited, in some regions diesel generation is problematic due to any combination of: a) air emission-related constraints and/or permitting requirements, b) restrictions on fuel storage, c) fire regulations, and d) limits on noise, etc. Use of biodiesel fuel may help to overcome some or most of those hurdles and to overcome NIMBY (not in my backyard) reactions.

6. GENERALIZING THE RANGE OF THE T&D DEFERRAL BENEFIT

Following is a generalized framework for estimating the financial benefit for deferral of an upgrade for one year using modular capacity additions, including MES. This example is an extension of the working example introduced in Section 1, involving an upgrade factor of 0.33 and a fixed charge rate of 0.11.

In Figure 14, there are six plots. They represent six MES power ratings – referred to as storage power in the figure. Importantly; storage power is a portion of the existing T&D equipment's capacity before the upgrade. As an example; if the original T&D capacity (before upgrade) is 12 MW, then 1% storage power indicates an MES whose power rating is 1% * 12 MW = 0.120 MW or 120 kW.

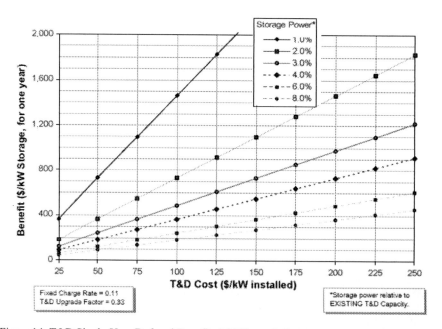

Figure 14. T&D Single-Year Deferral Benefit, 0.33 Upgrade Factor.

Note that this framework is *not* used to estimate storage system *size* (power or energy). Rather, it is used to estimate the *deferral benefit* for an MES with a given power rating and, presumably, with adequate discharge duration.

Consider the working example: using the preferred load growth estimate, distribution planners assume that load will exceed the 12 MW rating of the existing capacity by 250 kW. Power engineers specify an MES rated at 360kW (0.36 MW) to defer the upgrade at the site (storage power = 0.36 MW/12MW = .03 or 3%).

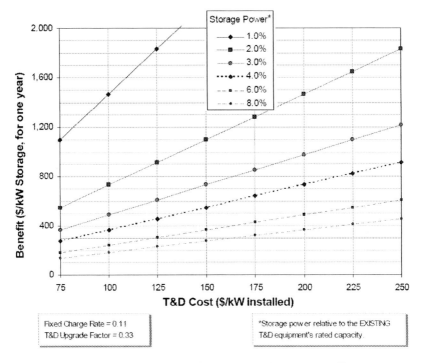

Figure 15. T&D Single-Year Deferral Benefit, 0.33 Upgrade Factor, Detail.

Consider again Figure 14 which reflects a T&D upgrade factor of 0.33 and the representative fixed charge rate of 0.11. If deferring a T&D upgrade which costs $75 per kW of *installed* capacity, the single-year deferral benefit is just under $350 per kW of MES, for one year. That is $350/kW for one year of T&D deferral, for MES rated at 360 kW. The generalized formula for calculating the annual T&D deferral benefit is:

(T&D installed cost * [1 + Upgrade Factor] * Fixed Charge Rate)/Storage Power

From this analysis, note that the most likely targets for deferral using modular resources are upgrades whose marginal cost is relatively high. That is because T&D upgrades with lower cost (on a per kW basis) are more likely to be cost-effective when compared to modular options. In Figure 14, the deferral benefits for projects with higher cost are plotted on the right side of the chart. Figure 15 emphasizes those values.

Figure 16 and Figure 17 show deferral benefit values for an upgrade factor of 0.25 and for an upgrade factor of 0.50, respectively. Note that the higher the upgrade factor, the higher the deferral benefit.

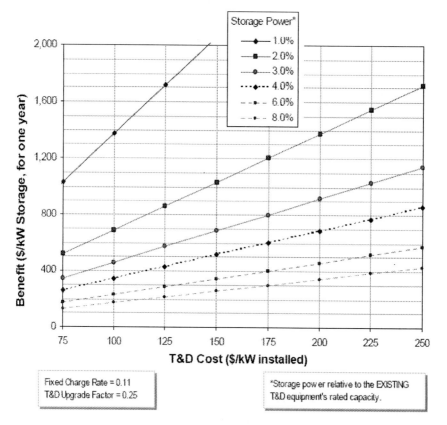

Figure 16. T&D Single-Year Deferral Benefit, 0.25 Upgrade Factor.

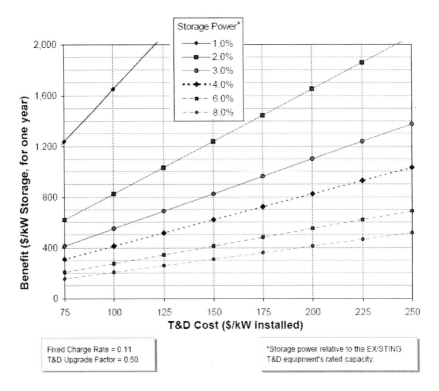

Figure 17. T&D Single-Year Deferral Benefit, 0.50 Upgrade Factor.

7. COMPLEMENTARY USES AND BENEFITS

This section describes uses of MES that could complement T&D deferral. That is important because, in many circumstances, it may be necessary to aggregate benefits associated with two or more applications – including T&D deferral – to comprise a value proposition whose combined benefits exceed the cost to own and to operate storage. In other words, there may be circumstances for which storage is not quite cost effective for T&D deferral alone, whereas combining benefits for one or more complementary uses may provide the extra value needed to justify use of MES when it would not be cost-effective for T&D deferral alone.

The degree to which uses provide complementary benefits (operationally and financially) depends on: 1) the coincidence between the power and energy needs for the complementary uses, 2) the degree to which additional MES use (beyond that needed for T&D-capacity applications) affects MES reliability, and 3) the incremental cost, if any, for MES enhancements needed for the MES to serve the complementary use(s).[6]

Introduction to Complementary Uses

A complementary use of MES with T&D deferral is one that:

1) does not conflict with use of the same MES for a T&D-capacity-related application (or conflicts can be managed or mitigated), and
2) adds value to the benefits derived from a primary use of MES for T&D deferral.

Consider this obvious, though important, premise: ideally, an MES or other DER used in lieu of T&D capacity must be available whenever load approaches the load carrying capacity (limit) of the T&D equipment. Possible consequences – if the MES fails to provide the expected level of service when needed – include reduced power quality, service outages, and T&D equipment damage. So, by definition, complementary uses must not reduce the MES's readiness to assist the T&D system when needed.

Depending on these criteria, and several possible institutional constraints, T&D deferral could be compatible with the following uses:

1) Wholesale Electric Energy Time-shift ("Buy Low – Sell High")
2) Electric Supply Capacity
3) Electric Supply Reserve Capacity
4) Reduce Transmission Congestion
5) Transmission Support and Stability
6) Voltage Support (Reactive Power)
7) Seasonal Deployment for Locational Benefits
8) On-Site Power Quality
9) Electricity Service Reliability
10) Retail Time-of-Use Energy Cost Reduction
11) Renewables Generation Firming
12) Renewable Energy Time-Shift
13) Demand Management and Curtailable Loads

Complementary Uses: General Considerations
Peak Demand Frequency

Typically, peak demand only occurs during a few days per year and for a few hours per day. Presumably, storage could be used for several other services during the many days and hours per year that it is not needed to serve the annual peak demand.

Power Rating

Some possible complementary uses of MES might require the MES to provide power (kW or MW) at the same time when the T&D system needs the MES's output. Some MESs can provide additional power – relative to their nominal rating – for short periods of time. Some of those systems are merely oversized for the primary application, but other MES types can be discharged at high rates, for short periods of time, usually with an efficiency and/or "wear and tear" penalty.

Stored Energy Limitation

Some possibly complementary uses of MES may compete for stored energy (kWh or MWh) such that there is not enough stored energy when T&D peak demand occurs. Depending on the type of MES, energy storage capacity can be added to an existing system.

MES Reliability

It should be noted that the discussion of complementary uses below does not address reliability- related effects, though it is certainly possible – if not likely – that complementary uses could reduce MES's reliability. Consider that, in general terms, increased use will probably increase wear and damage. For example; some types of storage (e.g. many battery types) are especially prone to performance degradation and premature failure if they are discharged too frequently and in some cases if they are discharged too much ("deep discharge") or not enough before being recharged.

"Perfect" MES Capacity

For this study, MES reliability is not addressed – MES is treated as if it is perfectly reliable. Of course no system is perfectly reliable. However, presumably, utility-owned and utility-rented MESs are well maintained and provide acceptable reliability. Depending on module sizes and the number of units involved, MES unit diversity may limit reliability concerns (i.e., if there are several smaller units, then the likelihood that multiple units will be out of service at any given time is low).

Regarding MES capacity that is not owned/rented and directly controlled by the utility: an important requirement for any non-utility MES used for capacity-related applications is "dispatchability" – storage output control via signals from external sources (e.g., from a location where multiple T&D nodes' operating conditions are monitored).

If MES is not owned by the utility and will serve load directly, then the necessary control could be accomplished indirectly by what is sometimes called "physical assurance". This involves a reduction of the amount of power that the utility delivers to the respective end-user site. The reduction is equal to the amount of MES capacity that the end-user has agreed to provide. In essence, the non-utility entity assumes the risk that the MES will serve as much load as expected, and will deliver the amount of energy expected.

Incremental Benefit/Cost Considerations

As described above, for some possible complementary uses, the MES power rating and/or energy storage capacity may have to be increased to concurrently accommodate both T&D deferral and complementary uses. If so, a simple benefit cost assessment will indicate whether the cost for more power and/or for more energy storage is commensurate with the added benefit.

Myriad Institutional Challenges

Because the concept of distributed energy resources (DERs) is evolving, there are many uncertainties about which DERs may be used and about how, where, by whom, and for how long. Each jurisdiction has its own perspective on the matter. Nonetheless, it is important to consider possible compatible uses without the encumbrance of the myriad challenges that might befall a specific project. Given this premise, the characterizations that follow *do not* address most institutional challenges. Instead these characterizations emphasize: a) operational compatibility/conflicts, and b) the degree to which combining benefits may affect the overall merits of a specific deferral.

Reactive Power Capability

Depending on the power conditioning equipment used, MES systems may be able to provide or to compensate for reactive power. (Please see Appendix C for more about reactive power.) That capability may make distributed MES more attractive. It can be used – within the distribution system – to offset undesirable effects caused by reactive loads. Reactive power from MESs could also be used in lieu of the ancillary service called voltage support. Both of those complementary uses are described in more detail below.

Complementary Uses

What follows are brief characterizations of the various uses of MES that are complementary to T&D deferral.

Electric Energy Time-shift
Charging MES with low-priced energy at night or on weekends and then discharging the energy when energy prices are high is a very compatible use (some refer to this application as "energy price arbitrage").

In this case, energy prices used are those for *wholesale* energy. That is, the energy price reflects the price of the electricity as a commodity, rather than reflecting the price paid by a retail energy user.

Consider these three important premises that affect the merits and viability of combining the electric energy time-shift benefit with the deferral benefit:

1) T&D capacity-related needs take precedent over the electric energy time-shift application. That is, the MES may only be used for "buy low – sell high transactions" if the MES will not be needed to serve in lieu of T&D capacity.
2) If it is worthwhile to buy low – sell high, there will be several hundred, perhaps thousands, of transactions per year. However, for many T&D capacity-related applications, the MES need only discharge for a relatively few hours annually (e.g., 0 to 200 hours per year).
3) The energy discharged when the MES is serving T&D capacity-related needs is likely to have substantial unit value, probably similar in magnitude to the value if the energy is discharged solely to optimize buy low – sell high profits.

So, the total benefits may be attractive when combining: 1) T&D-capacity-related benefits, 2) the value of the electric energy that is discharged when providing T&D capacity, and 3) electric energy time-shift benefits.

Electric Supply Capacity

MES used for localized T&D capacity-related needs may also provide a considerable benefit, related to the electric supply capacity, by reducing load that must be served by the electric supply and transmission system. The benefit can be substantial, depending on the degree to which T&D loading coincides with system loads.

Electric Supply Reserve Capacity

Reserve capacity is used when a power supply deficit materializes unexpectedly. Such emergencies can occur for several reasons. Common ones include major transmission circuits going out-of-service or becoming unstable, or a major plant having an unplanned shut down.

Two common reserve capacity categories are: 1) spinning reserve, and 2) supplemental reserve.

1) *Spinning Reserve* is electric supply capacity that can pick up load quickly. As the term implies, *spinning* reserve is normally comprised of rotating machinery (e.g., a turbine- generator set) that is actually spinning.
2) *Supplemental Reserve* is used during electric supply emergencies after all spinning reserve is in use. Supplemental reserve can include available generation that is currently operating at part load, rapid-start generation capacity, electricity storage capacity, or load management programs.

Many MES types can respond quickly enough that they could provide spinning-reserve-like service, by picking up load directly (to reduce load on the grid) or by providing power to the grid.

So, reserve capacity is a possible compatible application during times when: a) MES is not being used or will not be used soon in lieu of T&D capacity, and b) the MES has a sufficient amount of stored energy to provide capacity for the duration expected.

Interestingly, when the MES is being charged, it can be switched into discharge and provide roughly two times its rated output as reserve capacity. Consider a "very efficient" one MW system that is charged at about the same rate as it discharges. If the ISO needs to use reserves while the MES is charging, then: a) MES charging is halted, thus reducing load by about one MW, and b) the MES can pick-up load (by discharging), thus reducing demand on the supply system by another one MW. So, in effect, one MW of storage can provide two MW of reserve capacity while it is charging.

Reduce Transmission Congestion

As demand for electricity increases, transmission capacity additions are often not commensurate with growth. The result is that an increasing number of nodes in the transmission system are overloaded during an increasing number of hours per year, so all real-time demand cannot be served. To a lesser extent, congestion can cause transmission instability.

Storage can help. Most importantly, it can reduce the need for real-time transfer through congested transmission nodes. Instead, energy is transmitted and stored during non-peak demand periods, for use later when transmission is likely to be congested. Storage can also provide what might be called "transmission support" to reduce electrical effects from congestion, as described below.

Though *transmission* congestion receives most of the attention, congestion may also occur within an electricity *distribution* system, especially if there is a significant level of penetration by distributed generation on the distribution system.

Importantly, the primary financial benefit for reducing congestion is driven by the avoided cost for the conventional solution and/or the avoided cost to "do nothing." Assuming that the "do nothing" option is not acceptable, then, in essence, the financial benefit for reducing congestion is the avoided cost for a transmission upgrade – the T&D deferral benefit.

Transmission Support and Stability

Electric energy storage may be used to improve T&D performance by compensating for electrical anomalies and disturbances such as unstable voltage, voltage sag, and sub-synchronous resonance, to improve the system's performance and throughput. [7]

In the past, large power plants have been used to maintain stability. However, they tend to have a relatively slow response rate, so it is challenging for systems to respond precisely and quickly to rapid changes. Technological advances – such as those affecting communications and control, modern power electronics, and even superconducting materials – now make such transmission support technically viable.

Another facet of this use of storage involves attenuation of effects on the transmission system caused by high penetration of intermittent renewable energy generation, especially wind power. Benefits from transmission support are very situation-specific and site-specific, and there is no existing market for transmission support.[8]

As with reducing transmission congestion, the primary financial benefit for transmission support and stability is based on the avoided cost for the conventional solution and/or the avoided cost to "do nothing." If doing nothing is not acceptable, then the financial benefit is the avoided cost for a transmission upgrade – the T&D deferral benefit.

Power Factor Correction

Often, utilities use capacitors to compensate for effects from inductance, especially current that lags the grid's voltage. Such compensation is usually referred to as power factor correction. MES systems with reactive power capability could also be used for power factor correction with little or no conflict with use for T&D deferral.

Voltage Support

Voltage support – an ancillary service – is used to maintain the voltage of the electric supply and transmission system within a given region. Normally, reactive power needed for voltage support is provided by large central generation resources. Though voltage support is a relatively small portion of the total cost to generate and transmit all electricity, it does account for billions of dollars in total cost. Plus, newer central baseload generation technologies are not well-suited to use for reactive power generation. Conversely, new technologies – such as MES, modular generation, power electronics, and communications and control systems – make new alternatives possible.[9][10]

MES systems – especially those with reactive power capability – could be an attractive alternative to central generation voltage support, for several reasons. Perhaps most importantly, recent major power outages in the U.S. have been at least partially attributable to problems related to transmitting reactive power to load centers. Importantly, reactive power cannot be transmitted over long distances – like real power. So, voltage support is most effective when it is provided near to the loads that are most inclined to cause the voltage to drop or to stay low. Also, the conventional way to compensate for local voltage problems – capacitor banks – can actually exacerbate a system-wide voltage drop.

Depending on the location, if MES that is used for T&D deferral can provide reactive power, then the same MES could also provide reactive power, locally with little or no conflict. If the MES is charged and if it is not being used (to serve load for T&D deferral) then the MES could be used to serve a specific load so that the load is removed from the grid. That is especially helpful if the load served by the MES is one of the load types that affect voltage the most, such as small motors. Similarly, the MES could be used to inject real power into the grid. Doing either reduces the amount of power needed from the grid, which can reduce the need for voltage support.

On-Site Power Quality

Depending on its design and features, in many cases MES used for T&D deferral could also be used to keep important or high-value loads energized during times when power from the grid has reduced quality. Most power quality (PQ) related problems are caused by phenomena such as spikes, voltage sags or swells, sustained low voltage levels, frequency excursions and harmonics. Many PQ problems occur somewhat to very infrequently and/or have a short duration (ranging from milliseconds to minutes).

Existing solutions include, but are not limited to: a) PQ filtering equipment, b) more robust loads, and c) uninterruptible power supplies (UPSs) that filter, absorb, or otherwise compensate for power quality anomalies. UPSs range in size from those serving entire facilities to units serving specific loads (such as a personal computer).

Most PQ problems could be addressed using little or no energy from storage, and in most cases improving PQ would not conflict with use of the same MES for T&D deferral.

Electricity Service Reliability

Similar to power quality applications, electric service reliability involves the use of MES during service interruptions to: 1) keep important or high-value loads energized while the electric grid is not operational, or 2) allow for an orderly shutdown of equipment and processes, to minimize equipment and product damage and other losses. Service interruptions (also called outages) can last from a few seconds to many hours.

MES whose discharge duration is a few minutes to a few hours can be used to reduce negative effects associated with service outages. In some cases, MES with enough energy to serve load for a few minutes could be used like common UPSs: the MES serves the load for several minutes, allowing for an orderly shutdown of equipment and processes and possibly a smooth transition between utility power and other onsite backup power systems, such as emergency backup generators. In other cases, MES that has enough energy stored for one or more hours of service could be used for more sustained backup power service, perhaps in coordination with backup generation.

Similar to use of an MES for onsite power quality, use of the MES for reliability may be complementary to T&D-capacity-related uses. Also, like MES used for onsite PQ, an MES used for reliability and for T&D capacity applications must have capacity to store enough energy to both: a) satisfy T&D-capacity-related electricity needs, *and* b) carry load for the amount of time needed during service outages.

Consider an example: 1) an MES must discharge for four hours to reduce load on the T&D equipment and 2) critical and high value loads require about 20 minutes for an orderly shutdown. If allowing 20 + 10 = 30 minutes for the orderly shutdown, then an MES serves the T&D deferral and reliability applications must have enough energy to operate for a total of four hours and thirty minutes. The foregoing reflects the presumption that the MES will not be called upon to serve the T&D system and to serve on-site loads simultaneously.

Retail Time-of-Use Energy Cost Reduction

A potentially attractive use of MES is to reduce the retail electric energy bill for utility customers that pay time-specific prices for energy under provisions of a utility energy "time-of-use" rate structure. To reduce its electricity bill, a utility customer charges electricity storage when low (off-peak) energy prices apply, for use or even for sale to or via the grid when much higher on- peak energy prices prevail.

The discussion of the *wholesale* electric energy time-shift application as a complementary use (with T&D deferral) is germane: T&D capacity needs will take precedent, though T&D capacity needs are often coincident with times when energy is also quite valuable. So, depending on circumstances, using MES to reduce time-of-use energy cost may be somewhat, or even very, complementary to T&D capacity-related uses.

Consider an example: a utility seeks to reduce load so a T&D upgrade can be deferred. The utility has regulatory permission to pay customers downstream from the hot spot to reduce demand, up to twenty times in the next year, for as much as three hours per event. If that customer also qualifies for time-of-use energy pricing, then storage could be used to reduce the customer's total electricity cost by charging with off-peak energy for discharge when price is high. Presumably, weather forecasts would help storage owners and/or utilities gauge the likelihood that storage will be needed to support the T&D system.

Renewables Generation Firming

The topic of electricity storage to complement (firm) output from intermittent renewables is a topic of considerable, and growing, interest. The topic has increasing importance as interest in and use of renewables grows – to reduce air emissions, to improve energy security, and to reduce energy price risk and volatility.

There are no significant operational conflicts between use of the same storage system for T&D deferral and to store and optimize the value of energy from renewables. The most important consideration is location: for storage to enable T&D deferral, the storage must be electrically downstream from the T&D "hot spot."

If the renewable energy source and the storage are both downstream from the hot spot, then there should be no operational conflicts (assuming that the storage can be charged using electricity from the grid if energy from the renewable resource is not sufficient).

If the renewable energy source is located upstream from the hot spot, and the storage is downstream from the hot spot, then any operational conflicts should be manageable. That requires some combination of: a) selling renewable energy produced during peak periods directly to the electricity marketplace, real time, and b) "filling in" with off-peak grid energy, to charge storage, if renewable energy produced during non-peak periods is not sufficient to fully charge the storage.

So, storage used for T&D deferral may be compatible with storage of renewable energy (or potential conflicts can be managed) if: a) renewable energy may be stored when demand for utility power is below the load carrying capacity at the T&D hot spot, and/or b) storage is only needed for a few tens of hours to a few hundred hours per year for T&D deferral.

Renewable Energy Time-Shift

Storage can be used to store energy produced by renewable generation sources when the value of the energy is low, so that the energy can be used later, when it is more valuable. Note that this use of storage is distinct from renewable generation firming which emphasizes *ca*pacity/power benefits whereas energy time-shift emphasizes *energy/fuel related benefits*.

To some extent, storage used for renewable energy time-shift often encompasses elements of electric energy time-shift, especially if energy transactions – including those involving renewables – occur within the context of the normal electricity marketplace.

In many cases, time-shifting of renewable energy using storage could be somewhat or even quite complementary to use for renewable generation firming and to reduce renewable s-related transmission congestion.

Demand Management

Growing interest in load management and "demand response" as a resource is driven by its promise to support the electric supply system by: a) reducing demand for electricity that exceeds real-time supply, b) reducing congestion on the regional transmission system, and c) providing an alternative to reserve capacity (generation capacity that is held in reserve for use if major power sources or transmission become unavailable).

Energy storage could be an important element of electric supply-oriented demand management programs, either by picking up load or by providing electricity to the grid, when very high system demand, transmission congestion, or low reserve margins prevail.

With regard to T&D deferral: to the extent that the electric supply-related needs coincide with local peak demand (on the T&D equipment of interest, for deferral), the same storage could serve both the supply system needs and the localized T&D (capacity) needs.

Energy storage used for both supply and T&D deferral could: a) be owned and operated by utilities, and/or or b) enable end-users to participate in cost-effective load management by providing means to serve load if utility/grid service is unavailable or is "curtailed."

Seasonal Deployment for Locational Benefits

If an MES is transportable, it may be used in two (or more) locations within a utility if problems exist during different seasons. For example, a utility has two hot spots:

1) a circuit that is plagued by short periods of high peak demand, occurring for just a few days per year, during hot summer afternoons, driven by residential air conditioning loads, and
2) a circuit with dual peaks – morning and evening – driven by heating loads on cold winter days that cause heavy loading and voltage sags for a few hundred hours per year.

The key point is that the same equipment can provide more total benefits in a given year if it is used twice. In this case, the same equipment provides benefits in summer at one location, and then it provides additional benefits at the second location during winter.

Increased Flexibility

Flexibility is the degree to which and the rate at which adjustment to changing circumstances is possible. Flexibility may be an important source of enhanced value in a changing business environment, by enabling selection of a more optimal solution or response to a business related need or opportunity. However, estimating the value associated with flexibility may be challenging.

One helpful way to consider the value of flexibility is that it makes more optimal approaches possible. For example, MES or other types of modular resources used to provide T&D capacity on the margin – when and where needed – could provide more optimal T&D service than is possible with the conventional alternative (a normal T&D capacity upgrade).

Of course, depending on the circumstances, an optimal approach can involve criteria such as: a) higher revenue, b) more profit, c) lower cost, etc. When evaluating T&D deferral, the total cost for electric service may be the criterion of merit.

Two concepts can help: uncertainty and real options. Flexibility provides means to respond adeptly to uncertainty and thus to manage risk. Flexibility may provide ways to take advantage of business opportunities that are defined using real options.

Flexible Response to Uncertainty

Flexibility can be quite valuable when outcomes are subject to uncertainty. Uncertainty can be characterized as decision-makers' inability to estimate or predict a future outcome due to incomplete or imperfect knowledge about the spectrum and likelihood of possible outcomes.

The most important financial implication of uncertainty is the potential for increased financial risk. That financial risk usually manifests itself in the form of lower net financial returns than are required or expected. (Returns may be lower because revenues are lower than expected and/or because cost is higher than expected.)

Flexibility may allow managers to reduce or to limit downside risk (e.g., to limit reduced financial returns and higher financial losses). Flexibility may also provide bases for managers to accept risk – in an informed and prudent manner – to pursue additional upside potential (e.g., for increased financial net returns and/or reduced financial losses) for a given business endeavor.

Like all other businesses, electric utilities face uncertainty and related risk when making decisions that have financial implications, including decisions regarding investments in new or upgraded T&D capacity. For more details about managing T&D investment risk, see the next subsection of this report entitled Optimizing Risk-Adjusted Cost.

Evaluating Real Options

The concept of real options is a powerful way to consider some of the value of flexibility. Real options are potentially attractive business opportunities that could be pursued because a physical asset is currently owned or could be owned. They are referred to as *real* options because the possibility to use an asset actually exists. (Real options are different than the more familiar financial options: *financial* options exist in the form of a contract which conveys the right – but not an obligation – to purchase or to sell an asset, usually a financial asset like a stock or a commodity.)

Real options could also be characterized as "hidden" options because they remain unidentified by many asset owners who do not recognize or acknowledge them. Some real options are hidden (unrecognized) because they *could* exist if specific features or capabilities are added to an existing physical asset and/or if the asset owner can exploit operational or business synergies.

Generic real option examples could include: start a new project, develop or expand a business, narrow the scope of a business, postpone or abandon a project, or extend the time horizon for a project.

Consider an example scenario: the owner of an electricity storage system. The owner may have several real options, possibly including: do nothing, buy electric energy at a low price then sell the energy at a high price at the wholesale or retail level, provide ancillary services to the grid, or lease or rent the system to another user.

Importantly, real options are not alternatives in the sense that they are not different means to achieving a given goal or objective. Rather, real options are business activities that could be pursued because a physical asset is or could be owned.

Note that within the context of this report, modular resources are treated as an alternative to addressing the need to provide sufficient T&D capacity. Specifically, modular resources comprise one of three primary alternatives, the other two are: a) do nothing or b) do an upgrade.

Consider an example: utility-owned T&D capacity. Without considering regulation-related restrictions, real options related to utility-owned T&D equipment could include: a) do nothing (do not use the capacity), b) deliver electric energy, normally, c) use part of the T&D capacity to provide premium electric service, d) sell the equipment and real estate on which the equipment is located, and e) rent/lease some/all of the T&D capacity to third parties.

The possible purchase of additional T&D equipment to increase the T&D capacity could also enable the utility to: a) do nothing (do not increase T&D capacity), b) upgrade the T&D capacity for normal operations, or c) upgrade the capacity to take advantage of an opportunity to provide expanded and/or "premium" electric service.

Regarding modular resources that are or that could be owned by a utility: a real options perspective involves considering the potential to use modular resources to pursue various possible business endeavors. Such real options could include: a) do nothing (do not use the modular resource), b) rent or lease the equipment to another business, or one or more of the following: c) defer a T&D project and/or to defer the purchase of a larger central station generation, d) purchase low priced energy and sell that energy when price is high, e) provide ancillary services and f) increase local power quality or reliability.

Finally, by combining actual and possible utility ownership of T&D equipment and modular resources, the spectrum of real options includes a combination of the real options associated with the T&D capacity and those associated with modular resources separately. And, by virtue of being able to add various amounts of modular resources, there are various real options associated with various levels of modular resource deployment.

Compare that real option perspective to consideration of three primary response alternatives that could be used when load will exceed the load carrying capacity of the existing T&D equipment: a) do nothing, b) use modular capacity to serve load that exceeds the capacity of the existing T&D equipment, or c) do the standard T&D upgrade.

Optimizing Risk-Adjusted Cost

T&D planners face uncertainty when determining how much capacity is needed and when it is needed. Common sources of uncertainty include: a) the rate of normal load growth, b) the timing and disposition of block load additions, c) possible construction delays for scheduled upgrade projects (e.g., due to insufficient construction staff or capital, delayed permits), and/or d) the useful life remaining in T&D equipment. There may be others.

Traditionally, T&D planners have not had sophisticated enough planning tools or methodologies to include robust consideration of uncertainty during the T&D expansion planning process.

Uncertainty manifests itself as risk, an element of the *actual* cost of any business decision. T&D investments are no different. To the extent that T&D investment risk *does* manifest itself as an actual cost, that risk is spread among the utility's ratepayers. So, while the amount of risk incurred may be significant in any specific case or for any specific project, the amount is obscured because individual ratepayers pay a minimal amount to cover the total risk incurred throughout the service area.

Given this premise, using risk-adjusted cost as one key criterion for T&D investment decisions leads to lower overall cost-of-service when implemented across the utility's portfolio of T&D investments. Furthermore, to the extent that utility T&D capacity planners can evaluate uncertainty and risk robustly, they can manage or even accept or share risk when prudent and cost-effective. With emerging modeling and statistical techniques – such as predictive maintenance – accepting or sharing risk is possible, and in the future, it may even be commonplace.

One important reason that risk-adjusted cost was not used in the past is that when peak demand on T&D equipment approached the T&D equipment's load carrying limit, the two primary options available to the T&D planner included: a) upgrade the system, usually by adding a large increment of capacity, using conventional T&D equipment, or b) do nothing and hope that limits were not exceeded.

Now, when an upgrade is or will soon be imminent, T&D planners may include modular options such as MES in the evaluation for a much richer range of possibilities. Just one option is to use MES capacity to defer the need for an upgrade by serving marginal peak demand in the next year.

Other considerations also make this concept attractive as an important element of the value propositions for MES:

1) Though the risk-adjusted cost approach is a departure from traditional T&D expansion planning practices that are based on rules and reliability benchmarks:
    o The risk-adjusted cost approach has characteristics in common with existing T&D planning approaches, most notably is the need to address planning uncertainties and to accommodate MES effectively.
    o The risk-adjusted cost approach is also consistent with emerging T&D planning techniques that are more sophisticated, incorporating predicative maintenance, statistical modeling and other approaches to optimize T&D capacity use and life.

2) It seems important to consider more explicit and transparent treatment of risk as an important element of more sophisticated treatment of T&D (services) pricing.

A significant installed base of MES could also be an element of routine or even advanced electric supply and/or fuel-related risk management strategies.

8. CONCLUSIONS AND RESEARCH OPPORTUNITIES

In the future, modular energy resources, including modular electricity storage (MES), will be important elements of electric utilities' response to increasing: a) emphasis on environmental effects, b) competition, c) uncertainty, d) siting challenges, and e) challenges associated with integrating diverse "resources" (demand response, distributed renewables, other generation, etc.).

T&D upgrade deferral is an attractive opportunity to use MES to reduce utilities' electricity delivery-related cost. The opportunity is related to the inherent flexibility associated with modular resources – in this case, energy storage. With modular resources, utility T&D capacity planners have the means to manage and to optimize operations and to improve the overall efficacy of the electricity grid.

The key challenge for MES vendors and advocates is to identify and to quantify value propositions commensurate with MES's relatively high up-front cost. This requires skillful aggregation of technically viable and compatible benefits (benefits aggregation). In many cases, the most significant, or one of the most significant, benefits possible from MES is T&D upgrade deferral.

R&D Needs and Opportunities

The concepts described in this report – while based on sound principles – are new and largely untested. Also, T&D capacity planners and engineers, in general, have limited: a) authority, b) permission, c) tools, or d) familiarity/ experience required to apply these concepts or other new means to serve T&D capacity needs on the margin. So, the next steps in related research must present limited risk (e.g., equipment demos and experiments in the field), and the must address key knowledge and experience gaps.

Given these premises, there are several attractive, low-cost, low-risk opportunities for related research. It is important to note that, to one extent or another, these opportunities are being addressed in other technology realms, including: electric vehicles, plug-in hybrid vehicles, distributed photovoltaics and wind generation, demand response, "smart buildings," distributed generation, and "smart grids."

The author suggests a portfolio of four interrelated R&D thrusts to address important challenges affecting prospects for integrating the use of MES in standard T&D planning. The suggested portfolio also provides stakeholders with opportunities to learn more about how storage and other modular resources increase T&D planning options and flexibility, and how they may reduce utility cost-of-service, including risk. The four thrusts, described below, are:

- Standard Practice: Establish Criteria and Identify Next Steps
- Understanding Uncertainty and Risk
- Case Studies
- Enhanced Value Propositions: Best Prospect, PV Firming, and Grid Stability

Standard Practice: Establish Criteria and Identify Next Steps

An important research thrust is to establish a "roadmap" for developing formalized standards for the use of modular storage for T&D deferral. Two key near-term objectives include: 1) establish a catalog of stakeholders and of important stakeholder-specific decision criteria, and 2) identify key next steps.

This R&D thrust could include the scope of the Understanding Uncertainty and Risk thrust described below. It could also be a companion to, or part of, the Case Studies thrust also described below.

Understanding Uncertainty and Risk

A key challenge affecting the attractiveness of using MES for T&D deferral is that the uncertainties and risk associated with this approach are not yet well-characterized.

Utility T&D planners are quite familiar with the performance and reliability of the wires and transformers which dominate T&D systems. They have established, though evolving, means to accommodate consideration of service reliability when making T&D expansion decisions. Utility T&D planners will probably not use new or even evolutionary technology until it: a) is mandated, b) undergoes rigorous testing, and/or c) comes with solid warranties.

To address these challenges, an important objective is to characterize the ways and the extent to which use of MES and other modular resources can or will affect: a) electric service reliability and safety, b) grid operations, and c) performance of other T&D equipment.

So, an important R&D opportunity involves: a) characterizing evaluation criteria (needed to evaluate risk associated with use of MES for deferral), b) identifying and surveying existing and emerging sources for related information and data, and c) developing an initial, basic framework for evaluating and quantifying risk. [18]

Readers should note that there is significant potential to use MES to *reduce* the risk associated with investments in conventional T&D upgrades. Characterizing this element of the T&D deferral value proposition has important synergies with a characterization of ways that use of MES could *increase* risk. [14]

For example, MES used for T&D deferral increases the risk that T&D equipment will be overloaded if the storage is undersized. The same MES could reduce the risk that the utility will make an investment in an expensive upgrade to serve a housing development that is cancelled.

This R&D thrust could be part of the scope of the Standard Practice thrust, described above, or it could be a companion to, or part of, the Case Studies thrust, described below.

Case Studies

A low-cost, low-risk, high value research thrust is for planners to evaluate MES "on paper" for specific T&D expansion projects. Projects evaluated could be: a) actual T&D upgrades that are planned or that will be required within a few years, b) T&D upgrades that have already occurred, or c) "generic," synthetic, or model T&D upgrade situations.

The final product would be a compendium of cases and lessons learned.

Enhanced Value Propositions: Best Prospect, PV Firming, and Grid Stability

Though the T&D deferral benefit is significant in specific locations, it is likely that value propositions for MES used for T&D deferral will have to include other benefits to offset all costs for MES. Section 5 of this report lists benefits that could be compatible with T&D deferral. Two important, related R&D themes are:

1) Develop a first-generation decision framework/logic needed to "dispatch" MES so that total benefits are optimized when including T&D deferral and one or two other potentially significant benefits in a model value proposition. The one or two other benefits evaluated are presumed to be technically compatible.
2) Evaluate prospects for and characterize next steps needed to demonstrate two significant, mid-term benefits: a) capacity firming for distributed photovoltaics[15], and b) "critical system support" to stabilize the grid when area, region or system wide voltage collapse is possible.[1 6] [17]

TERMS USED IN THIS DOCUMENT

Application – A specific way or ways that energy storage is used, to satisfy a specific need; how/for what energy storage is used.
Arbitrage – See Bulk Electricity Price Arbitrage.
Avoided Cost – see T&D Avoided Cost.
Benefit – See Financial Benefit and Deferral Benefit.
Beneficiaries – Entities to whom financial benefits accrue due to use of a storage system.
Block Load Addition – An entirely new load that is to be connected to the grid. Examples include one-time load additions involving: a) new commercial and housing developments or b) new equipment with a large power draw relative to the load carrying capacity of the T&D equipment that serves the load.
Bulk Electricity Price Arbitrage (Arbitrage) – Purchase of inexpensive electricity during off- peak periods when demand for electricity is low, to charge the storage plant so that the low priced energy can be used or sold at a later time when demand/price for electricity is high.
Carrying Charges – The annual financial requirements needed to service debt or equity capital used to purchase and to install the storage plant, including tax effects. For utilities, this is the revenue requirement. See also Fixed Charge Rate.
Combined Benefits – Sum of all benefits that accrue due to use of an energy storage system, irrespective of the purpose for installing the system.
Complementary Benefit – a benefit that is complementary to a primary benefit because there are limited or no operational or technical conflicts.

DER – see Distributed Energy Resource. **Deferral Benefit** – see T&D Deferral Benefit.

Demand – The maximum power draw during a specific period of time, normally in units of kilowatts (kW) or megawatts (MW) for utilities (not adjusted for power factor).

Direct Cost – All direct costs to own or to rent an option, possibly including some or all of the following: rental charges, equipment purchase and delivery cost, project design, installation, depreciation, interest, dividends, taxes, service, consumables, fees and permits, and insurance.

Discharge Duration – Total amount of time that the storage plant can discharge, at its nameplate rating, without recharging. Nameplate rating is the nominal full load rating, not "emergency," "short duration," or "contingency" rating.

Discount Rate – The interest rate used to discount future cash flows to account for the time value of money. For this document the standard assumption value is 10%.

Dispatchability – Storage output control via signals from external sources (e.g., from a location where multiple T&D nodes' operating conditions are monitored).

Distributed Energy Resource – An electric resource (demand response, distributed generation, or energy storage) that is located at or near loads, usually within or at the end of the electricity distribution system.

Distribution – See Electricity Distribution. **Diversity** – See Unit Diversity.

Economic Benefit – The sum of all *financial* benefits that accrue to all beneficiaries using storage. For example, if the average *financial* benefit is $100 for 1 million storage users then the *economic* benefit is $100 * 1 million = $100 Million of *economic* benefit. See Financial Benefit.

Efficiency – The amount of energy that is discharged for each unit of energy used for charging.

Electricity Distribution – Electricity distribution is part of the electricity grid that delivers electricity to end-users. It is connected to the transmission system which, in turn, is connected to the electric supply system (generators). Relative to electricity *transmission,* the distribution system is used to send relatively small amounts of electricity over relatively short distances. In the U.S., distribution system operating voltages generally range from several hundred volts to 50kV (50,000 Volts). Typical power transfer capacities range from a few tens of MWs for substation transformers to tens of kilowatts for small circuits.

Electricity Subtransmission – As the name implies, subtransmission transfers smaller amounts of electricity, at lower operating voltages than transmission. For the purposes of this study, "transmission and distribution" is assumed to include subtransmission and not high capacity/high voltage transmission systems.

Electricity Transmission – Electricity transmission is the "backbone" of the electricity grid. Transmission wires, transformers, and control systems transfer electricity from supply sources (generation or electricity storage) to utility *distribution* systems. Relative to electricity *distribution* systems, the transmission system is used to send large amounts of electricity over relatively long distances. In the U.S., transmission system operating voltages generally range from 200 kV (200,000 Volts) to 500 kV. Transmission systems typically transfer the equivalent of 200 to 500 megawatts of power. Most transmission systems use alternating current though some larger, longer transmission corridors employ high voltage direct current.

Energy Density – The amount of energy that can be stored in a storage system with a given volume or mass.

Equipment Rating – The amount of power that can be delivered under specified conditions. The most basic rating is the "nameplate" rating: nominal power delivery rate under "design conditions." Other ratings may be used as well. For example, T&D equipment often has what is commonly called an "emergency" rating. That is the sustainable power delivery rate under "emergency conditions" such as when load exceeds nameplate rating by several percentage points. Operation at emergency rating is assumed to occur infrequently, if ever.

Financial Benefit (Benefit) – Monies received and/or cost avoided by a specific beneficiary, due to use of energy storage.

Financial Life – This is the plant life assumed when estimating lifecycle costs and benefits. A plant life of 10 years is assumed for lifecycle financial evaluations in this document (i.e. 10 years is the standard assumption value).

Fixed Charge Rate – The Fixed Charge Rate is used to convert capital plant installed cost into an annuity equivalent (payment) representing annual carrying charges for capital equipment. It includes consideration of interest and equity return rates, annual interest payments and return of debt principal, dividends and return of equity principal, income taxes, and property taxes. The standard assumption value for Fixed Charge Rate is 0.13 for utilities.

Inflation – The annual average rate at which the price of goods and services increases during a specific time period.

Inherent Load Growth – Routine or normal load growth mostly associated with increased business and leisure activities. Inherent load growth is also affected by effectiveness (or lack thereof) of energy efficiency and demand management programs.

Installed Cost – The cost to design, purchase and install the T&D equipment ($/kW nameplate of T&D equipment installed).

Lifecycle – See Financial Life.

Lifecycle Benefit – Present value of financial benefits that are expected to accrue over 10 years for a storage plant.

Marginal Cost – The cost to produce or to procure the next increment (e.g. of energy or capacity). The incremental cost is said to be the cost "on the margin."

MES – see Modular Electricity Storage.

Modular Electricity Storage – Electricity Storage that is or that can be deployed as numerous/many smaller modules rather than as one or a few large units.

Peak Demand – The maximum power draw on a power delivery system, usually year-specific.

Physical Assurance – Use of equipment that allows the utility to reduce power delivered to a specific end-user, under specific terms. It is used to ensure that end-users do not draw power beyond a certain rate during times when load is high. It is used for end-users that propose to provide peak capacity using distributed resources. An example of a formalized definition is "the application of devices and equipment that interrupt a DG customer's normal load when [a distributed resource] does not operate. (Derived from SDG&E Opening Brief-Phase 1, p. 31.)

Power Density – The power output per unit of volume or mass of a storage system.

Present Value – Present value is the total value of a series of payments and/or benefits expressed in terms of the value of money in a given year, normally the first year of an evaluation. Present value is calculated based on a) the value of payments and/or benefits in each of the periods evaluated and b) a specified discount rate. For example, for a discount rate of 10%, payments of $15 in year 1, $17 in year 2 and $23 in year three has a present value of $44.97, as shown below.

Year	1	2	3	Total
Value	15	17	23	**55**
Multiplier	0.909091	0.826446	0.751315	**2.486852**
Discounted Value	13.64	14.05	17.28	**44.97**

*Discount Rate = 10.0%.

Ramp Rate – The rate at which power output can change, whether up or down.

Real Options – Real options are potentially attractive business opportunities that could be pursued because a physical asset is currently owned or could be owned. They are referred to as real options because the possibility to use an asset actually exists.

Reserve Capacity – Generation capacity that is held in reserve for use if major power sources or transmission facilities become unavailable.

Revenue Requirement – For a utility, the amount of annual revenue required to pay carrying charges for capital equipment and to cover expenses including fuel and maintenance. See also Carrying Charges and Fixed Charge Rate.

Round Trip Efficiency – See Efficiency.

Spinning Reserve – Electric supply capacity – primarily generation but possibly storage – that is held in reserve and that can "pick up" load quickly. It is usually comprised of rotating machinery, such as a turbine-generator set, that is actually spinning.

Standby Losses – Energy losses that occur when the storage is charged but not being used (e.g., for pumped hydroelectric it is water evaporation in the upper reservoir, for CAES it is loss of air pressure due to air escaping from the reservoir).

Supplemental Reserve – Electric supply capacity used during electric supply emergencies after all spinning reserve is in use. It may include available generation that is currently operating at part load, rapid-start generation capacity, electricity storage capacity, or load management programs.

Storage Discharge Duration – See Discharge Duration.

Storage Power Rating - Storage power rating is the amount of storage capacity (kW) needed relative to the existing T&D capacity.

Subtransmission – See Electricity Subtransmission.

Transmission and Distribution (T&D) Deferral – Delaying an investment in and construction of new or upgraded T&D equipment.

T&D Avoided Cost – The cost not incurred by utility ratepayers if the T&D upgrade is not made. The avoided cost is equal to the revenue requirement.

T&D Deferral Benefit – The cost that will not be incurred (cost that is avoided) if a given T&D project upgrade is deferred. For utilities, that amount is the annual revenue requirement: the amount of money that must be collected from utility ratepayers at large to cover the single-year cost.

Unit Diversity – Using several or many units rather than one or a few units to address a given need. This is done to increase reliability by reducing the likelihood that a catastrophic outage will occur (because it is less likely that several or many smaller units will fail when compared to one or a few larger units).

Value Proposition – A value proposition is comprised of all benefits and all costs, including risk, that are associated with an investment or purchase.

REFERENCES

[1] Hadley, S. W. Van Dyke, J. W. Poore, W. P, III. Stovall, T. K. *Quantitative Assessment of Distributed Energy Resource Benefits*. Oak Ridge National Laboratories. Report ID: ORNL/TM-2003/20. May 2003. Page 35.

[2] *Focus On Distribution System Regulation: Avoiding Costs And Capturing Values*. The Rate Assistance Program. Issuesletter. February 2003. http://www.raponline.org.

[3] Energy and Environmental Economics, Inc., and Rocky Mountain Institute, *Methodology and Forecast of Long-Term Avoided Costs for the Evaluation of California Energy Efficiency Programs*, Prepared for: California Public Utilities Commission Energy Division, October 25, 2004.

[4] Woo, C. Lloyd-Zannetti, D. Orans, R. Horii, B. (Energy and Environmental Economics) and Heffner, G. (EPRI). *Marginal Capacity Costs of Electricity Distribution and Demand for Distributed Generation*. The Energy Journal. 1995.

[5] Pupp, Roger. *Distributed Utility Penetration Study*. Sponsored and published by Electric Power Research Institute (EPRI). EPRI Report TR-106265. March 1996.

[6] Eyer, James. Iannucci, Joe. Corey, Garth. *Energy Storage Benefits and Market Analysis Handbook, A Study for the DOE Energy Storage Systems Program.* Sandia National Laboratories. Report# SAND2004-6 177. December 2004. http://www.sandia.gov/ess/.

[7] Mears, D. Gotschall, H. *EPRI-DOE Handbook of Energy Storage for Transmission and Distribution Applications.* Report # 1001834. December 2003.

[8] Behnke, Michael R. Erdman, William L. *Impact of Past, Present and Future Wind Turbine Technologies on Transmission System Operation and Performance.* Prepared for the California Energy Commission. Report Number CEC-500-2006-050. March 9, 2006.

[9] Li, F. Fran. Kueck, John. Rizy, Tom. King, Tom. *Evaluation of Distributed Energy Resources for Reactive Power Supply, First Quarterly Report for Fiscal Year 2006.* Prepared for: U.S. Department of Energy by Oak Ridge National Laboratory and Energetics Incorporated. November 8, 2005.

[10] Kirby, Brendan. Hirst, Eric. *Ancillary Service Details: Voltage Control.* Oak Ridge National Laboratory. Energy Division. Sponsored by The National Regulatory Research Institute. Report ORNL/CON-453. December 1997.

[11] Eyer, James. Iannucci, Joe. E*stimating Electricity Storage Power Rating and Discharge Duration for Utility Transmission and Distribution Deferral, A Study for the DOE Energy Storage Program.* Sandia National Laboratories, Energy Storage Program, Office of Electric Transmission and Distribution, U.S. Department of Energy. November, 2005. Sandia report #SAND2005-7069. http://www.sandia.gov/ess/.

[12] IEEE 519-1992 is the relevant standard established by the Electrical and Electronics Engineers, Inc. http://www.ieee.org/web/standards/home. http://www-ppd.fnal.gov/ EEDOffice-w/Projects/CMS/LVPS/mg/ 8803 PD9402.pdf.

[13] Eyer, James. Iannucci, Joe. *Estimating Electricity Storage Power Rating and Discharge Duration for Utility Transmission and Distribution Deferral, A Study for the DOE Energy Storage Program.* Sandia National Laboratories, Energy Storage Program, Office of Electric Transmission and Distribution, U.S. Department of Energy. November 2005. Sandia report #SAND2005-7069.

[14] Eyer, James. Hoff, Thomas. Iannucci, Joe. Pupp, Roger. Skeen, Jim. *Comparing Electric T&D Capacity Options, Including Stationary and*

Transportable Distributed Energy Resources, On a Risk-adjusted Cost Basis. Draft Report. June 2006. http://dua.jimeyer.net/docs/MESRisk.pdf.

[15] Hoff, Thomas E. Perez, Richard. Margolis, Robert M. *Maximizing the Value of Customer-Sited PV Systems Using Storage and Controls.* Presented at the American Solar Energy Society (ASES) 2005 Conference. http://www.clean-power.com/research/ customerPV/Outage Protection_ASES_2005 .pdf.

[16] Eromon, David I. Kucck, John. Rizy, Tom. *Distributed Energy Resource (DER) Using FACTS, STATCOM, SVC and Synchronous condensers for Dynamic Systems Control of VAR.* 2005 National Association of Industrial Technology (NAIT) Convention. Saint Louis. November 16-19, 2005.

[17] Mozina, Charles. *Undervoltage Load Shedding – Parts 1 and 2.* Electric Energy T&D Magazine. May 2006. http://www.electricenergy online. com/.

[18] Eyer, Jim. Iannucci, Joe. Pupp, Roger. Skeen, James. Hoff, Thomas. *Comparing Electric T&D Capacity Options, Including Stationary and Transportable Distributed Energy Resources, On a Risk-adjusted Cost Basis.* A draft report for Sandia National Laboratories. March 2009.

APPENDIX A. MES TRANSPORTABILITY EXAMPLE DETAILS

What follows is the worksheet used to generate the values charted in Figure 3 in Section 2. That figure shows financials for transportable DER capacity. Values plotted are shown in the last row of the worksheet.

Option: Modular Storage

	Max Load Growth	"Over-sizing"**	Capacity
Modular Storage Sizing	240	50%	360
		*for engineering contingency	

Discount Rate 10.0%
Cost Escalation 2.0%
D Cost Annualization Factor** 0.11

**Used to calculate "annuity" value that is treated as annual cost-of-ownership (or revenue requirement); fixed costs, including financing cost, taxes, and interest.

Current Dollar values reflect the *actual* amount paid in the respective year, expressed in terms of the value of money in the respective year.

Real Dollar values reflect the price to be paid in the respective year *without* including projected cost escalation.

Present Worth (PW) values reflect a "discounting" of current dollar values, by the discount rate, to express the value of money recieved at a future time in terms of what that money is worth today, if that money were used for the "next best" investment.

Year #	1	2	3	4	5	6	7	8	9	10
Load Growth Rate	2.0%		2.0%		2.0%		2.0%		2.0%	
Existing D Capacity (MW)	12		12		12		12		12	
D Capacity Added (MW)	4		4		4		4		4	
Upgrade Project Cost ($000 $Real)	1,200		1,200		1,200		1,200		1,200	
Upgrade Project Cost ($000 $Current)	1,200		1,248		1,299		1,351		1,406	
($/kW$_{Installed}$, $Current)	75		78		81		84		88	
($/kW$_{Added}$, $Current)	300		312		325		338		351	
"Other" Benefit** ($000 $Real)	27.0		27.0		27.0		27.0		27.0	
"Other" Benefit ($000 $Current)	27.5		28.7		29.8		31.0		32.3	
**e.g., PQ and Reliability										

Single Year Cost-of-ownership for Upgrade ($)

Load Growth, in respective year (kW)	240	n/a	240	n/a	240	n/a	240	n/a	240	n/a
Total										
$Current 715,526	132,000	0	137,333	0	142,881	0	148,653	0	154,659	0
$PW 499,148	132,000	0	113,498	0	97,590	0	83,911	0	72,150	0

Single Year "Other" Benefit ($)

Total										
$Current 149,285	0	27,540	0	28,653	0	29,810	0	31,015	0	32,267
$PW 94,673	0	25,036	0	21,527	0	18,510	0	15,915	0	13,685

Single Year Benefit ($)

Total										
$Current 864,811	132,000	27,540	137,333	28,653	142,881	29,810	148,653	31,015	154,659	32,267
$PW 593,822	132,000	25,036	113,498	21,527	97,590	18,510	83,911	15,915	72,150	13,685

Modular Storage Value ($/kW PW)

Modular Storage Capacity (kW) 360

Total										
$/kW-yr, $Current 2,402	367	77	381	80	397	83	413	86	430	90
$/kW-yr, $PW 1,650	367	70	315	60	271	51	233	44	200	38

APPENDIX B. STORAGE FOR T&D DEFERRAL – TWO CASE STUDIES

Pacificorp Vanadium Redox Batteries

Situation

Utah Power, a subsidiary of Scottish Power-affiliate PacifiCorp, owns and operates a 25 kV distribution circuit serving about 11 MVA of load in several small communities, along the circuit's 209 miles (129 miles of the total is a 3-phase circuit).[B1]

The Rattlesnake #22 feeder is located in an environmentally sensitive, undeveloped area in the Moab region of Southeast Utah, near Arches National Park. The feeder's right-of-way follows the Colorado River Valley.

Unacceptable reliability and power quality problems led to repeated complaints by utility customers to the Public Service Commission. Serving additional customers was likely to reduce line voltage to unacceptable levels during peak demand periods.

Existing mitigation measures include seven reclosers, five step-type voltage regulators, two 300 kVAR fixed-tap capacitor banks. Several additional mitigation options were considered, including: electricity storage, substation upgrades, circuit upgrades, and increased compensation for reactive power.

Solution

The option selected was VRB Power Systems' (VRB) patented "vanadium redox battery" electricity storage system with an "intelligent four-quadrant power converter." The system – described by the vendor as a "regenerative fuel cell" – has continuous power output of 250 kW and can store 2,000 kWh for an eight-hour discharge duration.

The power conditioning unit can provide 250 kVAR of reactive power support and has a true power rating of 353 kVA. The system is housed in a simple 350 square meter metal building in Castle Valley, Utah. The building has room to double the system's output.

The VRB system was selected based on its overall benefit/cost relationship and, to a lesser extent, because: a) it had a relatively short lead time, b) institutional challenges, especially environment-related, were surmountable, c) the modular system is readily upgraded, to add as much as twice the original capacity, and d) facility construction is straightforward.

One notable consideration: generation was ruled out, in large part, because of concerns about possible negative effects on power quality and voltage from generation "without selective or dynamic control of the power injected."

Regarding the cost/benefit: the deferral benefit alone is significant. In addition, the facility was designed for unattended operation which translates into lower cost. Also, environment-related challenges and costs, including permitting, were manageable. Notably, the inherent characteristics of the VRB technology allowed for a straightforward permitting process, even in this environmentally sensitive area, especially when compared to challenges that would face a circuit upgrade. Also, just the *option* to double output from the current facility – at low incremental cost – is valuable, as it provides an important hedge against load growth uncertainty.

Service

The system was installed in November 2003 and was formally commissioned in March 2004. It relieves congestion on the feeder by: a) reducing the load on the circuit during peak demand periods, using energy stored when demand is low, and b) providing "rapid response" voltage support.

In late 2004, the vendor "enhanced system efficiency and reliability through the upgrade of the power conversion system and cell stacks."

The system is normally controlled by an on-site algorithm, though it can be controlled remotely. Among other important controlled parameters: the system's power output is changed in a way that allows slower voltage regulators to keep pace with related voltage changes. Additional improvements are expected with respect to voltage support as operators gain experience with the system's automated VAR support capability.

Status

The vendor has reported that the system "has provided full power daily cycling operations since March, 2004 and has effectively served its intended purpose through a full summer peak season." Among other important results, improved power factor reduced line losses to the extent that the circuit's load carrying capacity increased by as much as 40 kW. That, according to the vendor, offsets the losses that are inherent to the energy storage.

The vendor reports that Utah Power is evaluating the merits of and timing of a possible expansion of the system, to provide "additional capacity and reliability enhancements to the community."

AEP Sodium/Sulfur Battery Situation

Appalachian Power Company, an operating unit of American Electric Power (AEP), determined that a substation located near Charleston, West Virginia required an expensive upgrade to serve growing peak demand.[B2] Based on an informal survey of eight selected AEP T&D projects, installed costs ranged from $26/kW to $ 169/kW, for an average of about $74/kW ($2003). [B3]

Solution
In September 2005, AEP committed to use a 1.2MW / 7.2 MWh hour electricity storage system employing sodium/sulfur (Na/S) battery technology to defer the substation upgrade by six to seven years, beginning in summer 2006.

The Na/S system is similar to a smaller Na/S battery system that has been in operation at AEP offices near Columbus, Ohio, since 2002 to serve demand and power-quality needs.[B4]

An important project partner is NGK Insulators Ltd. (NGK) whose NAS Battery Division provided the battery. Another important partner is S&C Electric Company's Power Quality Products Division which provided the power electronics and system integration. Also, the U.S. Department of Energy (DOE) is a supporting sponsor through Sandia National Laboratories (SNL).

Service
Energy stored in state-of-the-art Na/S batteries serves summer daytime peak demand, reducing maximum power draw on the substation equipment. After service at the initial location, expected to be several years, AEP may opt to move the system to another location.

Project partners expect the battery to last 4,000 to 5,000 charge-discharge cycles at 90% depthof-discharge, or as long as fifteen years. Notably, according to NGK, they have commercialized the Na/S battery in Japan, in concert with the Tokyo Electric Power Company. As of September 2005, NGK had over 125 MW / 750 MWh in service.

According to AEP, the Na/S system "represents an exciting step into the future" that "should provide valuable information about potential uses elsewhere in the AEP system" and that is expected to help AEP "use its distribution, transmission and generation assets more efficiently." Deployment of the system "is consistent with AEP's short-term objective" which is "to deploy storage systems selectively, based on energy cost savings and on where the systems can defer upgrades to our distribution system without compromising safety or reliability."

Status

The following is an excerpt from a report published by Sandia National Laboratories regarding the financial benefits from the demonstration.

> The economic analysis shows a rate of return of 9.8%, which exceeds the 7% discount rate. Under the technical and economic assumptions described, the system represents an economically favorable investment because it provides a return greater than the cost of money. The present value of power quality (PQ) benefits was estimated at $791,000, compared with peak shaving (PS) benefits of $217,000. Thus, PQ benefits represent 78% of the total benefits, PS only 22%. PQ benefits are site specific, however, and finding places with high PQ payback requires knowledge of specific sectors and participants. While PS benefits are not as large, it would be relatively easy to find utilities that offer tariffs more favorable to PS operation than the 'typical' tariff assumed here. Overall, the results suggest that the dual application of the NAS® BESS does provide potentially attractive economics. The feasibility of specific projects, however, must use actual cost data, estimates of customer-specific avoided outage costs, and the actual terms of the local utility tariff. [B4]

Since that report, APS has issued press releases indicating a growing interest in batteries. Over the longer term AEP may install as much as 1 GW of MES to "boost reliability and help to integrate wind generation."[B5]

According to the press release, AEP expects to install 6 MW of stationary sodium Na/S battery technology in its 11-state service territory during 2008 and 2009. It will initially install units in West Virginia and Ohio... [in 2008] and will work with wind developers to identify a third location for battery technology deployment. As of late 2007, plans were to "add two megawatts of NAS battery capacity near Milton, W.Va., to enhance reliability and allow for continued load growth in that area" and to "add two megawatts of NAS battery capacity near Findlay, Ohio, to enhance reliability, provide support for weak sub-transmission systems and avoid equipment overload."[B5]

Appendix B. References

B1. Use of VRB Energy Storage System for Capital Deferment, Enhanced Voltage Control and Power Quality on a Rural Distribution Feeder Utility – A Case Study in Utility Network Planning Alternatives. VRB Power Systems, Incorporated. February 28, 2006. http://www.vrbpower.com.

B2. Press Release: AEP'S Appalachian Power Unit to Install First U.S. Use of Commercial- Scale Energy Storage Technology. http://sandia.gov/ess/About/ docs/Press_9-19- 05_AEP.pdf.

B3. Eckroad, Steve. Nichols, Dave. Utility-Scale Application of Sodium Sulfur Battery. The Battcon International Stationary Battery Conference, March 2003. http://www.battcon.com/PapersFinal2003 /NicholsPaper FINAL2003 .pdf.

B4. Norris, Benjamin L. Newmiller, Jeff. Peek, Georgianne. NAS® Battery Demonstration at American Electric Power. A Study for the DOE Energy Storage Program. Sandia National Laboratories. Report# SAND2006-6740. March 2007. http://www.sandia.gov/ess/.

B5. Press Release: AEP to deploy additional large-scale batteries on distribution grid installations will boost reliability, integrate wind generation, prepare for future; new batteries a step toward AEP's goal of 1,000 megawatts of advanced storage. Ohio, Sept. 11, 2007. http://www.aep.com/newsroom/newsreleases/default.asp?dbcommand= displayrelease&ID= 1397.

APPENDIX C. REACTANCE AND POWER FACTOR

Electrical characteristics of power grids and of electric loads can lead to undesirable voltage (level) variations due to electrical phenomena called "reactance." Reactance occurs because equipment that generates, transmits and utilizes electricity may exhibit characteristics like those of inductors and capacitors in an electric circuit.

In simple terms, for electric power systems that use the alternating current (AC) form of electricity, capacitors and capacitance cause current to "lead" the system's voltage. Conversely, inductors and inductance cause current to "lag" the voltage.

To the extent that reactance does cause the current to lead or to lag the voltage, the magnitude of the voltage is lower than the intended level. That, in turn, reduces the amount of load that the power system can serve.

Note that, in electricity power grids, reactance from inductors and inductance is far more common than reactance from capacitors and capacitance, leading to current that lags the voltage as shown in Figure C-1 below. Common inductive loads are: a) motors, b) compressors, c) lighting ballasts, and d) any electricity-using equipment that includes voltage transformers.

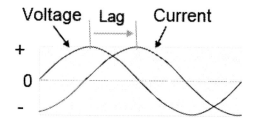

Figure C-1. Current Lag Due to Inductance in an Alternating Current Power System.

Reactance is often characterized by citing the ratio of real power to apparent power (i.e., the "power factor"). If current and voltage are exactly synchronized then the system has what is called a "unity" power factor equal to 1.0. (To the extent that current *lags* voltage the power factor is less than one.)

The key effect in the distribution system is that the presence of reactance reduces the amount of real power – usable power – that can be delivered to end users. For example, equipment rated at 480 Volts and 500 Amps has an "apparent power" of 480 * 500 = 240 kiloVolt-Amps (kVA). That is also the amount of real power that can be delivered if there is no reactance. But, if there *is* reactance then real power is less than 240 kVA. For example, if the power factor is 0.85 then the real power delivered is 240 kVA * 0.85 = 204 kiloWatts.

Utilities use various means to compensate for the presence of reactance in the distribution system – primarily due to inductive loads – to maintain voltage. Perhaps the most common way is installation of capacitors. Because most reactance is caused by inductive loads, current in T&D systems tends to *lag* the voltage. Furthermore, because capacitors cause current to lead the voltage, they can be used to offset effects from inductors.

End Notes

[1] Consider an example: A 12 MW T&D node will be upgraded so it can serve more load. If the upgrade factor is 0.25, then 3 MW will be added, for a total of 15 MW. If the upgrade factor is 0.50, then 6 MW will be added, for a total of 18 MW.

[2] For the purpose of this discussion, units of true power (kiloWatts or MegaWatts) and apparent power (units of kiloVolt-Amps or MegaVolt-Amps) are used interchangeably. In practice, there are important technical and cost differences. In short, various load types reduce the effectiveness of the grid by, for example, injecting harmonic currents or by increasing "reactance." As a general indication of the difference as it affects utility distribution systems, consider this example: a power system serves 10 MW of peak load (true power). During times when load peaks the power factor may drop to 0.85, so the T&D equipment should have an apparent power rating of at least 10 MW/0.85 = 11.76 MVA.

[3] As context: for a 360kW storage system the PQ/reliability benefit is 360kW * $75/kW = $27,000 per year. That is $27,000/12MW = $2.25 per year, per kW of peak demand served by the T&D equipment. In that example, the removed 12 MW transformer has a residual value of: 15 years remaining/40 years life * $3 0/kW= $11.25/kW ($135,000)

In: Modular Electricity Storage
Eds: B.N. Mendell and L.P. Brunwick

ISBN: 978-1-61470-459-1
© 2012 Nova Science Publishers, Inc.

Chapter 2

BENEFIT/COST FRAMEWORK FOR EVALUATING MODULAR ENERGY STORAGE[*]

Susan M. Schoenung and Jim Eyer

NOTICE

Prepared by
Sandia National Laboratories
Albuquerque, New Mexico 87185 and Livermore, California 94550
 Sandia is a multiprogram laboratory operated by Sandia Corporation, a Lockheed Martin Company, for the United States Department of Energy's National Nuclear Security Administration under Contract DE-AC04-94AL85000.
 Approved for public release; further dissemination unlimited.

Issued by Sandia National Laboratories, operated for the United States Department of Energy by Sandia Corporation.
 This report was prepared as an account of work sponsored by an agency of the United States Government. Neither the United States Government, nor any agency thereof, nor any of their employees, nor any of their contractors, subcontractors, or their employees, make any warranty, express or implied, or

---

[*] This is an edited, reformatted and augmented version of a Study for the DOE Energy Storage Systems Program publication, Sandia Report SAN D2008-0978, dated February 2008.

assume any legal liability or responsibility for the accuracy, completeness, or usefulness of any information, apparatus, product, or process disclosed, or represent that its use would not infringe privately owned rights. Reference herein to any specific commercial product, process, or service by trade name, trademark, manufacturer, or otherwise, does not necessarily constitute or imply its endorsement, recommendation, or favoring by the United States Government, any agency thereof, or any of their contractors or subcontractors. The views and opinions expressed herein do not necessarily state or reflect those of the United States Government, any agency thereof, or any of their contractors.

Printed in the United States of America. This report has been reproduced directly from the best available copy.

ABSTRACT

The work documented in this report represents another step in the ongoing investigation of innovative and potentially attractive value propositions for electricity storage by the United States Department of Energy (DOE) and Sandia National Laboratories (SNL) Energy Storage Systems (ES S) Program. This study uses updated cost and performance information for modular energy storage (ME S) developed for this study to evaluate four prospective value propositions for MES. The four potentially attractive value propositions are defined by a combination of well- known benefits that are associated with electricity generation, delivery, and use. The value propositions evaluated are: 1) transportable MES for electric utility transmission and distribution (T&D) equipment upgrade deferral and for improving local power quality, each in alternating years, 2) improving local power quality only, in all years, 3) electric utility T&D deferral in year 1, followed by electricity price arbitrage in following years; plus a generation capacity credit in all years, and 4) electric utility end-user cost management during times when peak and critical peak pricing prevail.

ACKNOWLEDGMENTS

This work has been sponsored by the United States Department of Energy (DOE) Energy Storage Systems (ES S) Program under contract to Sandia National Laboratories (SNL). The authors would like to thank Imre Gyuk of DOE and Paul Butler, John Boyes, and Nancy Clark of SNL for their support.

SNL is a multiprogram laboratory operated by Sandia Corporation, a Lockheed Martin Company, for the United States Department of Energy's National Nuclear Security Administration under Contract DE-AC04-94AL85000.

ACRONYMS AND ABBREVIATIONS

CAES	compressed air energy storage
CPP	critical peak pricing
DER	distributed energy resource
DG	distributed generation
DOE	Department of Energy
DUA	Distributed Utility Associates
ESS	energy storage system
I^2R	resistive losses (current squared * resistance)
kV	kiloVolt
kVA	kiloVolt-Amps
kW	kiloWatt
kWhout	kiloWatt-hours output
Li-ion	lithium-ion
MES	modular energy storage
MW	megawatts
Na/S	sodium/sulfur
Ni/Cd	nickel/cadmium
O&M	operation and maintenance
PCS	power conversion system
PG&E	Pacific Gas and Electric Company
PW	present worth
PQ	power quality
SNL	Sandia National Laboratories
T&D	transmission and distribution
UPS	uninterruptible power supply
VAR	volt-amp reactive
V-redox	vanadium-redox
VRLA	valve-regulated lead-acid
Zn/Br	zinc/bromine

CONVENTIONS USED IN THIS REPORT

For simplicity, units of power, or load carrying capacity, will be expressed in units of kilowatts (kW), although in some cases units of kiloVolt-Amps (kVA) may be more appropriate. For example, utility equipment is rated in units of kVA rather than kilowatts. For the purposes of this study, the distinction is not important.

The term transmission and distribution (T&D) is used throughout this document. It is important to note that the focus of this study is on distribution and subtransmission systems, rather than the higher voltage, higher capacity, "bulk" transmission systems. Two key reasons for this are: a) criteria used to decide whether to add transmission capacity are somewhat different than those used to justify a subtransmission or distribution upgrade, and b) the roles for distributed energy resources (DERs) that serve the transmission system directly (e.g., to stabilize voltage or frequency) may be different than the roles served by DER used for subtransmission and distribution capacity (i.e., used in lieu of actually transmitting energy). So, in this report, the term T&D refers to subtransmission and distribution.

EXECUTIVE SUMMARY

Purpose

The work documented in this report was undertaken for three key purposes:

1) Often, benefits and costs developed in previous energy storage studies were computed using different financial bases. This work reconciles those financial bases so that costs and benefits are expressed using consistent bases and assumptions.
2) The Energy Storage Systems (ESS) Program management at Sandia wanted to update their storage technology cost and performance information to reflect state-of-the-art.
3) Results in this report reflect another next step in the ongoing investigation of innovative and potentially attractive value propositions for electricity storage by the Department of Energy (DOE) and Sandia National Laboratories (SNL) ESS Program.

Scope

The scope of this report covers:

1) Characterization of a basic framework for evaluating the benefits and costs for modular energy storage (MES) that is used for various applications. The framework includes common financial bases and consistent assumptions for both cost and benefit calculations.
2) Up-to-date MES system cost and performance data for ten leading electricity storage technologies.
3) Estimates of MES costs and benefits for four, possibly attractive electric utility-related value propositions.

Intended Audience

The intended audience for this report includes utilities and electricity providers (planners, engineering, and management), electricity storage vendors, technology developers, system integrators and advocates, and energy policymakers and researchers.

Key Results and Conclusions

Of the ten technologies considered, lead-acid batteries appear to have the greatest potential for attractive benefit / cost ratios in the combined T&D deferral / power quality value proposition. In general, to improve the benefit / cost ratio for all cases, costs for energy storage systems must be reduced. Opportunities for combining benefit values have the greatest potential to result in attractive benefit/cost ratios for all technologies.

1. INTRODUCTION

This work combines results from previously separate research sponsored by the U.S. Department of Energy (DOE) Energy Storage Systems (ESS) Program at Sandia National Labs (SNL). A primary objective is to establish a framework for expressing electricity storage benefits and costs using consistent assumptions and bases. The framework is exercised using up-to-date cost and performance projections for leading modular energy storage (MES) technologies.

The evaluation compares costs and benefits for four, potentially attractive uses of MES (value propositions). They involve use of MES for:

Value Proposition 1: transportable MES for electric utility transmission and distribution (T&D) equipment upgrade deferral in even numbered years and for improving local power quality in odd numbered years, at different locations.

Value Proposition 2: transportable MES for improving local power quality in all years, at different locations.

Value Proposition 3: electric utility T&D deferral in year 1, followed by electricity price arbitrage in following years; plus a generation capacity credit in all years.

Value Proposition 4: electric utility end-user cost management during times when peak and critical peak pricing prevail.

The value propositions evaluated include financial benefits identified in previous work by Distributed Utility Associates (DUA) sponsored by the DOE ESS Program [1] and the California Energy Commission [2], and using costs based on previous work by Longitude 122 West and Advanced Energy Analysis for the DOE ESS Program that were updated for this study [3, 4]. This work is a follow-on to related analyses performed previously for the ESS Program [5, 6].

2. METHODOLOGY DESCRIPTION

Financial Assumptions

One objective of this study was to establish generic criteria for calculating energy storage benefits and costs, using consistent bases and assumptions. This section describes the approach and assumptions used for financial analysis.

Readers should note that two key assumptions – storage system service life and the discount rate used to calculate present worth over the service life – are intended to represent a generic circumstance. For any specific circumstance, other more situation-specific assumptions may be appropriate. A general indication of the effect that service life and discount rate have on lifecycle financials is provided later.

Benefits and costs associated with storage system use are calculated using common financial bases, shown in Table 1. Most notable, in order of significance, are: a) ten year storage system service life, b) 10% discount rate, and c) 2.5% annual price escalation (inflation) rate.

Table 1. Assumptions for Life Cycle Benefit and Cost Analysis

Parameter	Value
Service life	10 years
Discount rate	10%
General inflation rate	2.5%
Utility Fixed Charge Rate	11%
Fuel cost, natural gas (for surface CAES only)	5 $/MBTU
Electricity cost, charging	5 ¢/kWh

Those three criteria – service life, discount rate, and inflation – are used to calculate the Present Worth (PW) Factor. The PW Factor provides a simplified way to represent a discounted present worth of a stream of regular revenues or payments, for a given number of years.

Present Worth Factor

Consider a simple example as an illustration of how the PW factor is used. In year 1 of a project, the total cost is $1.00. In subsequent years, that annually recurring cost is assumed to escalate at 2.5% per year, due to inflation. The upper plot in Figure 1 indicates annual cost as it escalates from year to year. The values plotted are referred to as "current dollar" values.

The lower plot in Figure 1 indicates the present worth of the current dollar values (shown in the upper plot), after applying the discount rate for the respective number of years. Those "discounted" values are summed over all years of the project to calculate the present worth for all years. Note the dramatic impact that discounting at 10% per year has on the final value.

The curves in Figure 1 show that, for a cost of $1 in the first year, after ten years of inflation at 2.5%/year, the cost in year ten would be about $1.28 (current dollar value). Discounting that same $1 (in year ten) at 10%/year results in a present worth or discounted value of about 50 cents (for year ten).

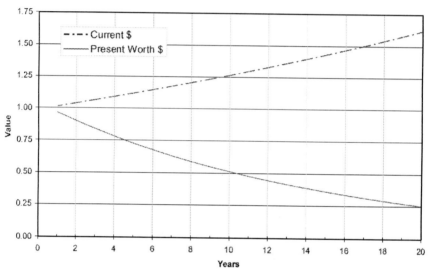

Escalation Rate: 2.5%/yr., Discount Rate: 10.0%/yr.

Figure 1. Annual Current and Present Worth Values for $1 in Year 1.

Figure 2 shows cumulative values for the same two series shown in Figure 1. That is, they are the cumulative values for escalated cost and for discounted annual values. For a given year, the value of the upper plot represents the cumulative amount reflecting a cost of $1 in year 1, escalating at 2.5%/year (current dollar value). The lower plot represents the cumulative amount when summing annual present worth (discounted) values reflecting a 10% discount rate.

Figure 2 indicates that, after ten years, the cumulative value of costs incurred annually – beginning with $1 in year 1 and escalating at 2.5% each year – is about $12.00. When discounting that amount for ten years, the resulting present worth is about $7.17. That reflects a Present Worth Factor (PW factor) of 7.17 for a project lasting ten years if inflation is 2.5%/year and the discount rate is 10%/year.

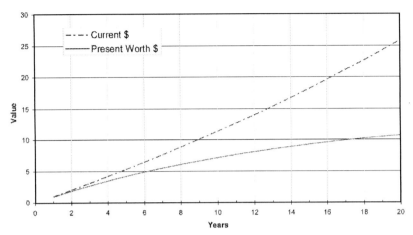

Escalation Rate: 2.5%/yr., Discount Rate: 10.0%/yr.

Figure 2. Cumulative Current and Present Worth Values for $1 in Year 1.

The equation for the PW factor for a ten-year service life is as follows:

$$\text{PW factor} = \sum_{i=1}^{10} \frac{(1+e)^{i-.5}}{(1+d)^{i-.5}}$$

e = annual price escalation rate (%/year)
d = discount rate (%/year)
i = year

Figure 3 shows PW factors for three discount rates, assuming a cost escalation of 2.5%/year. (The value of "i" is calculated at mid-year.) For a given life/discount rate combination, the PW factor represents the present worth for a stream of values like that described in Figure 2. Note that the plot for the 10% discount rate in Figure 3 represents the same values as the lower plot in Figure 2.

Figure 3 allows for quick comparisons of annually recurring costs and benefits for various project service lives and discount rates.

Consider another example. Assume that a storage plant will cost $100,000 in the first year of operation. That annual cost is expected to escalate at 2.5% per year over the ten year service life. The owner uses a 10% discount rate. The present worth of all costs (before tax) is about $717,000 (7.17 PW factor *

$100,000 in year 1). For comparison, look at the outcomes for the other discount rates in Figure 3. For a first year cost of $100,000, the present worth (over ten years) is about $813,000 if the discount rate is 7%/year and the ten year present worth is about $630,000 if the discount rate is 13%/year.

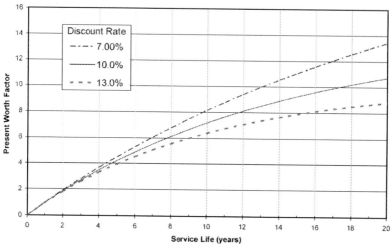

Escalation Rate: 2.5%/yr. Mid Year Convention.

Figure 3. Present Worth Factors for various Service Lives and Discount Rates.

3. VALUE PROPOSITIONS

A value proposition is comprised of all benefits and all costs, including risk, that are associated with an investment or purchase. Four value propositions are used to illustrate the benefit/cost evaluation framework described in this report. Those value propositions were selected because they involve opportunities for which MES is technically viable and could yield high benefits. Many other value propositions are possible and may show advantages for storage technologies.

>Value Proposition 1: transportable MES for electric utility transmission and distribution (T&D) equipment upgrade deferral in even numbered years and for improving local power quality in odd numbered years, at different locations.
>Value Proposition 2: transportable MES for improving local power quality

in all years, at different locations.

Value Proposition 3: electric utility T&D deferral in year 1, followed by electricity price arbitrage in following years; plus a generation capacity credit in all years.

Value Proposition 4: electric utility end-user cost management during times when peak and critical peak pricing prevail

The assumed operating and design parameters for the four value propositions are listed in Table 2.

4. STORAGE TECHNOLOGY

Storage Options Evaluated

The energy storage technologies evaluated for each value proposition are listed in Table 3. The characteristics of these technologies are derived from previous work found in References 3, 4, and 7.

For value proposition 1, the storage technologies evaluated are those which are movable and which can provide storage for five hours of discharge. Though flywheels are probably not suitable for applications requiring hours of storage, they are suitable for power quality/reliability applications, and are therefore included in the value proposition 1 analysis. Surface Compressed Air Energy Storage (CAES) is evaluated only for value proposition 3, where a moveable resource is not required.

Storage Cost and Performance

The cost approach is the same as that in Reference 3. The cost and performance data for the various storage technologies were derived based primarily on Reference 3, although more up-to- date information was also used. Notably, the authors found limited changes during the years 2005 and 2006, especially with regard to equipment (system) installed cost.

Specific exceptions to Reference 3 include:

- Replacement costs were reduced for storage components used for value proposition 1 in this study because the number of operating cycles is far less than the five full discharge / recharge cycles per week that were assumed in the previous work.

Table 2. Parameters of Value Propositions for Energy Storage Benefit / Cost Analysis

	Value Proposition 1: Transportable MES for T&D Deferral and PQ	Value Proposition 2: Transportable MES for improving PQ	Value Proposition 3: T&D Deferral Plus Energy Price Arbitrage	Value Proposition 4: Peak Plus Critical Peak Electricity Pricing
Description	1st year deferral, 2nd yr PQ/reliability; move to new location; 3rd year deferral, 4th year PQ, etc.	10 years all power quality	1 year deferral, subsequent years arbitrage	Operate during peak and critical peak hours to avoid time- of-day charges and earn discount
Power range	300 kW – 1 MW	300 kW – 1 MW	500 kW – 2 MW	20 kW – 1 MW
Hours of dispatchable	4 - 5 hrs	0.25 – 1 hr	4 – 5 hrs	5 hrs
Hours of operation per year	T&D: 200 hrs/yr PQ: 20 hrs/yr	PQ: 20 hrs/yr	T&D: 200 hrs/yr Arbitrage: 1000 hrs/yr	Critical peak: 60 hrs/yr Total peak: 1600 hrs/yr
Technology issues	Must be moveable, suitable for infrequent use, rapid availability		Routine use, high duty cycle	Routine use, high duty cycle

Table 3. Storage Technologies Evaluated

Value Proposition 1: Transportable MES for T&D Deferral and PQ	Value Proposition 2: Transportable MES for improving PQ	Value Proposition 3: T&D Deferral Plus Arbitrage	Value Proposition 4: Peak Plus Critical Peak Electricity Pricing
• Lead-acid batteries (flooded and VRLA) • Ni/Cd • Na/S batteries • Li-ion batteries • Zn/Br batteries • V-redox batteries • High-speed and low-speed flywheels • Lead-carbon asymmetric caps • Hydrogen fuel cell	• Lead-acid batteries (flooded and VRLA) • Ni/Cd • Li-ion batteries • Zn/Br batteries • High-speed and low-speed flywheels • Lead-carbon asymmetric caps	• Lead-acid batteries (flooded and VRLA) • Na/S batteries • Ni/Cd • Li-ion batteries • Zn/Br batteries • V-redox batteries • Surface CAES • Lead-carbon asymmetric caps • Hydrogen fuel cell	• Lead-acid batteries (flooded and VRLA) • Ni/Cd • Na/S batteries • Li-ion batteries • Zn/Br batteries • V-redox batteries • Lead-carbon asymmetric caps • Hydrogen fuel cell

- Vanadium-redox battery costs have dropped: for this study, energy-related equipment cost is $350/kWh, compared with $600/kWh in Ref. 3. [8]
- Lead-carbon asymmetric capacitors were not included in Ref. 3, but have been added based on Ref. 9 and 10. The assumed costs are $500/kWh for the energy-related equipment component and $350/kW for the power-related equipment.

Thus, the capital cost assumptions for storage technologies considered in this study are shown in Table 4. Balance of Plant includes the auxiliary components outside of the storage subsystem or power converters. For some technologies, these costs are integral to the power system.

Table 4. Storage Technologies Evaluated and Costs

Technology	Energy- Related Cost ($/kWh)	Power–Related Cost ($/kW)	Balance of Plant ($/kW)
Lead-acid Batteries (Flooded Cell)	150	175	50
Lead-acid Batteries (VRLA)	200	175	50
Ni/Cd	600	175	50
Zn/Br	400	175	0
Na/S	250	150	0
Li-Ion	500	175	0
V-redox	350	175	30
Technology	Energy- Related Cost ($/kWh)	Power–Related Cost ($/kW)	Balance of Plant ($/kW)
Lead-carbon asymmetric caps	500	350	50
CAES-surface	120	550	50
High-speed flywheel	1,000	300	0
Low-speed flywheel	380	280	0
Hydrogen fuel cell	15	1500	0
Electrolyzer (to accompany fuel cell)	None	300	None

As a possibly helpful comparison, consider the costs for state-of-the-art small uninterruptible power supplies (UPSs) with ratings ranging from 700 VA to 30 kVA shown in Table 5, below.

Table 5. State-of-the-art UPS Ratings and Prices

Device	VA	Minutes	Price ($)	$/kVA	$/kVA-hour
LIEBERT - UPSTATION GXT2 700VA UPS- RM 2U 17MIN 4RCPTL	700	17	601	858	3,028
TRIPP LITE - SMART OL 1000VA RM TWR 2U-UPS 6RCPTL USB SER SLOT	1,000	6	443	443	4,430
MGE UPS SYSTEMS INC - PULSAR EX RT 3200VA RM-OL RACK/TWR UPS	3,200	6	1,130	353	3,530
Tripp Lite SmartOnLine SU7500RT3U - UPS - 6 kW - 7500 VA	7,500	9	3,017	402	2,682
APC - SMARTUPS VT OL 30KVA RM 3-PH 208V 5 BATT PDU STRTU	30,000	13.7	27,000	900	3,942

Notes:
- This table does not constitute an endorsement or recommendation of the listed products or suppliers.
- Power rating in units of Volt-Amps.
- Typically 1.2 to 1.3 Volt-Amps are required for each Watt of load.
(source: www.techonweb.com)

5. RESULTS: BENEFIT – COST COMPARISONS

What follows is a summary of results and key assumptions. The results are benefit/cost values for the various MES options that were evaluated for the respective value propositions. Additional details are provided for value proposition 1, to illustrate the framework used to derive the results.

Value Proposition 1: Transportable MES for T&D Deferral and Power Quality

This value proposition involves transportable MES that is used in odd numbered years for high or relatively high value utility T&D upgrade deferral, and then moved for use in even years to improve localized power quality or reliability, over its ten year life.

A 5-hr energy storage system is assumed. Figure 4 indicates the operation hours assumed for each of the ten years of operation.

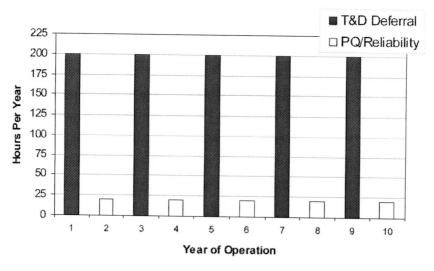

Figure 4. T&D Deferral and Power Quality: Hours of Operation per Year.

Benefit

The financial benefit for T&D upgrade deferral is assumed to last for one year at a given location, after which it becomes cost-effective to proceed with the T&D upgrade. That is a conservative assumption because MES may allow for multi-year deferrals.

The financial benefit for improving power quality and reliability arises when the MES is used to a) shield loads from electrical effects caused by short duration power quality events, and b) pick up load during long duration power outages, to ride though the outage or to allow for an orderly shut-down.

The annual benefit ($Current) for each of ten years is shown in Figure 5. Those values are estimated by assuming a benefit of $250/kW T&D deferral ($Constant) and a power quality benefit of $75/kW ($Constant). [10]

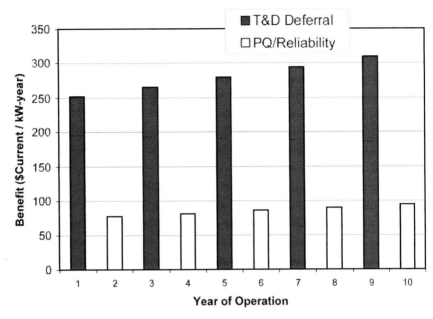

Figure 5. T&D Deferral and Power Quality: Benefits ($Current/kW).

Cost

As an example, a representative MES system is used to illustrate the cost breakdown. The representative MES is a state-of-the-art flooded lead-acid battery whose roundtrip efficiency is 75%. The costs to own and operate that system are shown in Figure 6.

Cost elements include installed capital equipment, maintenance, battery replacement, and the cost for charging energy. The replacement cost is significantly reduced from the previous study because the system is cycled much less frequently.

As noted in Section 4, MES used for T&D deferral and PQ is cycled much less frequently than was assumed in the original study, resulting in a proportional reduction in annual replacement costs. In the original cost study, the replacement period for the energy storage component of the system was assumed to be six years, and a complete cell replacement was assumed. The system was assumed to operate 5 days per week, 50 weeks per year, or a total 1,250 hours per year for a 5-hour system. In this current scenario, however, the system is cycled infrequently.

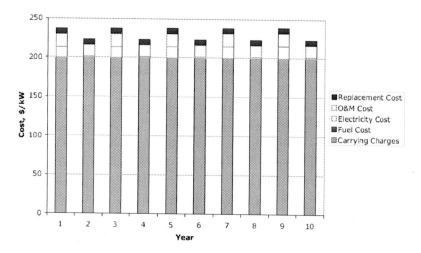

Figure 6. T&D Deferral and Power Quality: Current-year Cost Components for a flooded Lead-Acid Battery System ($/kW).

Benefit/Cost Results

Figure 7 shows the present value of a) benefits, and b) costs for the representative storage type. When applying the generic financial assumptions, present worth of benefits over 10 years is $1,203/kW. The 10-year present worth of cost, assuming the lower annual replacement cost is $1,200/kW. Thus, if replacement costs can be reduced as indicated above, it is possible to achieve a benefit / cost ratio slightly greater than one.

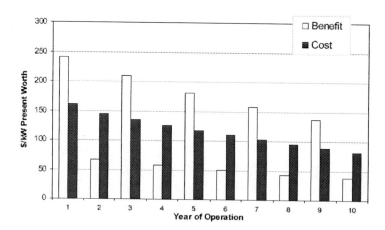

Figure 7. T&D Deferral and Power Quality: Present Worth of Benefits and Costs ($/kW) for a flooded Lead-Acid Battery System.

Other technologies are technically viable for value proposition 1. Based on costs from the previous study – with an adjustment of the replacement cost to account for reduced cycling – the present worth of benefits and costs over 10 years are shown in Figure 8 for all storage technologies evaluated. Note that lead-acid batteries have the lowest cost and thus have the highest benefit/cost ratio.

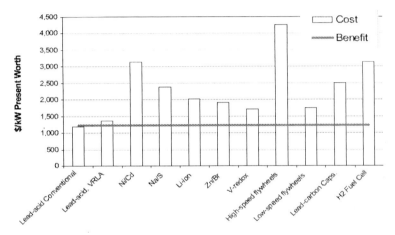

Figure 8. T&D Deferral and Power Quality: Present Worth of Benefits and Costs over 10 years.

Value Proposition 2: Improving Power Quality/Reliability

This value proposition involves use of MES for localized or on-site power quality improvement and/or to improve localized electric service reliability. Improving PQ and reliability is an especially attractive use of MES because: a) the relatively modest amount of storage (discharge duration) required, and b) the small number of lifetime discharges required, especially full discharge. That is especially true for MES technologies that have a high incremental cost for energy storage, i.e., long discharge duration.

In this case, a somewhat conservative discharge duration of 15 minutes was assumed. This is conservative in the sense that in many cases much less energy storage is needed.

Benefit

The financial benefits for improving power quality and reliability arise when the MES is used to: a) shield loads from electrical effects caused by short duration power quality events, and b) pick up load during long duration power outages, to ride though the outage or to allow for an orderly shut-down.

The annual benefit ($Current) for power quality for each of ten years is the same as in value proposition 1: $75/kW ($Constant).

Benefit/Cost Results

As shown in Figure 9, most MES types have attractive benefit / cost ratios for PQ/reliability applications.

Value Proposition 3: T&D Deferral Plus Energy Price Arbitrage and Central Generation Capacity Credit

For this value proposition, an energy storage system with five hours of discharge duration is deployed to defer a T&D upgrade for one year. After the first year, the system remains at the original location and is used for energy price arbitrage. It also receives a generation capacity credit equal to the annual cost avoided for additional central generation capacity in each of ten years.

T&D Deferral

For this value proposition, a "very high" T&D deferral benefit of $650/kW (of storage) is assumed for the first year of MES operation, which is the only year for which the T&D deferral benefit accrues.

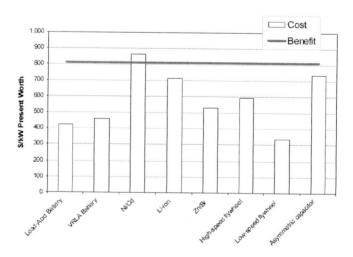

Figure 9. MES Technologies' Benefits and Costs for Power Quality.

That amount represents the highest 10% (most valuable) opportunity for T&D deferral. [2] Discharge operation for 200 hours per year (at full load equivalent) is assumed for T&D deferral.

Energy Price Arbitrage

Electric energy price arbitrage exploits wholesale electricity price volatility and diurnal variability. Benefits accrue when storage is charged using low-priced off-peak energy, for sale when energy price is high.

Storage losses or inefficiencies add to the fully burdened discharge cost. Fully burdened incremental discharge cost includes cost for charging energy, energy lost during the charge/discharge cycle, plus variable operating cost. Variable operating cost is comprised of the incremental cost for regular maintenance and the incremental cost for replacement of the storage medium, for each kWh discharged. So, for a given transaction, the net benefit (e.g. in units of ¢/kWhout) is:

Sell Price - (Charging Energy Price ÷ Storage Efficiency) - O&M - Replacement Cost.

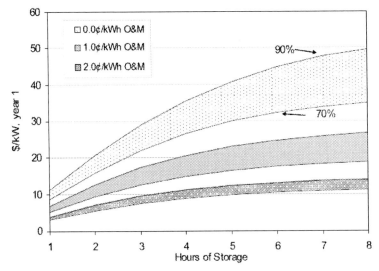

Figure 10. Net Arbitrage Benefits, $/kW.

The annual net benefit assumed for arbitrage was evaluated using a simple model developed by DUA [1] that uses "perfect knowledge" about future prices to "look ahead," to identify profitable sell opportunities. The dataset assumptions – comprised of 8,760 hourly price points, one for each hour of the year – is an electric energy price projection for California that was created by a production cost model.

Results shown in Figure 10 reflect net benefit for storage operated for one year. That is, they reflect the benefit remaining after accounting for the incremental cost for charging energy, energy losses, and variable operating cost (variable O&M and replacement costs).

Results are provided for storage systems whose variable operating cost (O&M and replacement cost per $_{kWhout}$) ranges from: a) nothing, to b) 1 $_{¢/kWhout}$, and c) 2¢/kWh$_{out}$. The data band shown for each cost scenario reflects net benefits for storage efficiencies ranging from 70% to 90% with storage

Central Generation Capacity Credit

The central generation capacity credit is estimated based on the premise that storage will be discharging during or close to peak demand periods and assuming a benefit that is based on the cost for a new peaker power plant. Based on the costs for several types of combustion turbine power plants, an installed cost of $5 00/kW is assumed. [12] That installed cost is used with the 0.11 fixed charge rate described in Section 2 to estimate the annual first year benefit of $500 * 0.11 = $55/kW-year. Over ten years, that is a present worth of 7.17 * $55/kW-year = $395/kW.

Combined Benefits

The benefits for value proposition 3 are indicated in Figure 11.

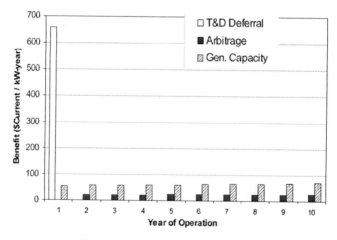

Figure 11. Present Worth of Benefits for Value Proposition 3.

Benefit/Cost Results

Annual benefit/cost results – reflecting cost and performance of the aforementioned state-of-the-art flooded lead-acid battery system – are shown in Figure 12.

Ten-year benefit / cost results for all storage technologies evaluated are shown in Figure 13. The state-of-the-art flooded lead-acid battery provides the highest benefit / cost, primarily because that technology has the lowest capital equipment cost.

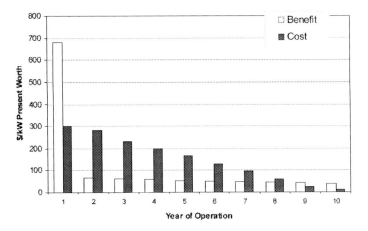

Figure 12. T&D Deferral, Arbitrage and Capacity Credit: Present Worth of Benefits and Cost for a flooded Lead-Acid Battery System.

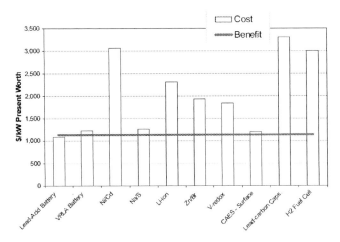

Figure 13. T&D Deferral, Arbitrage and Capacity Credit: 10-year Present Worth of Benefits and Costs for Various Storage Technology Types.

Value Proposition 4: Peak and Critical Peak Electricity Pricing

For this value proposition, a commercial electricity end-user's utility cost is reduced by using storage during peak and critical peak times. As of late 2005, Pacific Gas and Electric Company (PG&E) was offering Critical Peak Pricing (CPP) to some customers. In this situation, a customer is offered a discount on electricity prices during "non-critical" times (e.g. when generation reserve margins are adequate). This offer is made if the customer agrees to pay "very high" prices during critical peak periods in return for the discount. These prices can be as high as 5 times the normal peak energy charge, and be expected to prevail several times per year (the PG&E target is 12 times/yr), for periods lasting from 3 to 6 hours per event. Thus, a customer might make such an agreement using energy storage.

Benefit

Critical peak and peak pricing benefits were calculated based on PG&E E19, Medium Commercial Rates. [11] The annual benefit due to CPP provisions for operation of 12 full 5-hr discharges per year (60 hrs) is estimated as $25/kW per year. The annual benefit for operating storage during all peak-period price hours – including demand charge reduction – is $100/kW per year. 1,600 hours per year of storage discharge is assumed. These values have been used to calculate benefits for the CPP value proposition. The annual benefits are shown in Figure 14.

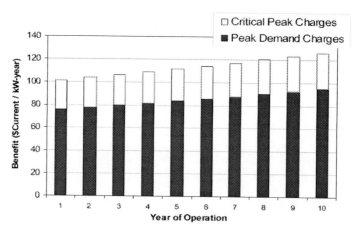

Figure 14. Peak and Critical Peak Pricing Value Proposition; Annual Benefits ($Current).

Benefit/Cost Results

Results (present worth of benefits and costs) for a flooded lead-acid battery system operating in value proposition 4 are shown in Figure 15, for each of 10 years. Even using off-peak electricity costs, the benefits are insufficient to break even. As seen in Figure 16, none of the technologies evaluated has a cost that is commensurate with the benefit.

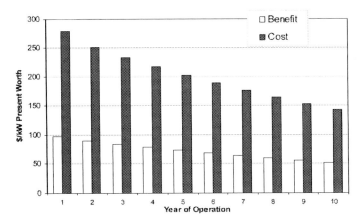

Figure 15. Peak and Critical Peak Pricing Value Proposition: Present Worth of Annual Benefits and Costs for 5-hr flooded Lead-Acid Battery.

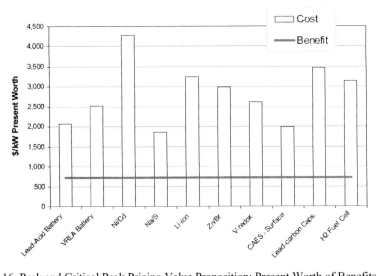

Figure 16. Peak and Critical Peak Pricing Value Proposition: Present Worth of Benefits and Costs for Storage Technologies, 10-year Lifecycle.

6. CONCLUSIONS AND RECOMMENDATIONS

Summary Results

Figure 17 shows the present worth of benefits and costs for lead-acid battery storage used for the four value propositions investigated. Most notably: value proposition 1 (T&D deferral plus PQ), value proposition 2 (PQ/reliability only) and value proposition 3 (high value T&D deferral plus arbitrage and generation capacity credit) show promise as they have a benefit/cost ratio greater than 1.

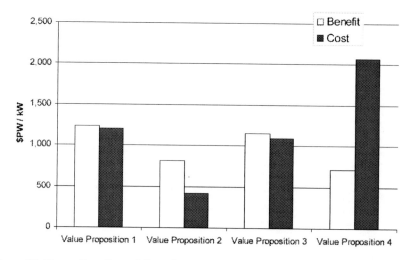

Figure 17. 10-year Benefits and Costs for a flooded Lead-Acid Battery System for Value Propositions 1, 2, 3 and 4.

CONCLUSION

Four value propositions were evaluated using an analysis approach that compares costs and benefits on a common basis. Other value propositions may also be defined to show potential viable storage applications. In this study, given existing and expected price signals and the authors' estimates of benefits associated with value propositions 1, 3, and 4, the costs for most types of MES are somewhat or significantly higher than estimated benefits. An exception is the case of short duration storage for power quality only (value proposition 2).

Regarding the arbitrage benefit, results suggests that energy storage system variable operating cost – especially maintenance and replacement cost – must be quite low for arbitrage to be attractive. The challenge is that, even for regions with "high" energy prices, during most hours in the year, the wholesale energy price is determined by large power plants with low marginal cost. Those prices are just too low for profitable arbitraging during most hours of the year unless the variable operation cost (not related to charging energy) for storage is also low. Specifically, as shown in Figure 10, if the non-energy variable operating cost (O&M plus replacement) exceeds 2¢/kWh$_{out}$, then the annual net benefit is likely to be quite low. Importantly, the non-energy variable operating cost, even for the most cost-effective MES technologies, is likely to exceed 2¢/kWhout for the foreseeable future.

Given the low benefit for arbitrage (relative to storage cost), the best prospects for energy storage – especially modular, distributed storage – are value propositions involving use of the MES as "capacity resources" that offset the need for other capital equipment.

For some energy storage technologies – those for which each discharge results in relatively significant equipment degradation, and thus high replacement costs – the best value propositions are those involving relatively few discharges.

Perhaps the best overall example of a good capacity-related value proposition is MES used to defer T&D upgrades. MES could also be used to augment the T&D system (e.g. for increased reliability and/or improved power quality), in lieu of other equipment, temporarily or permanently.

Of course, storage used for local capacity-related needs (i.e. T&D deferral or to improve power quality) can also provide regional benefits to the grid, including reduced demand for generation and transmission capacity, voltage support, reduced T&D I^{2}R energy losses, rapid response operating reserves, etc.

That leads to the most important conclusion of this study: the most compelling value propositions for modular energy storage will likely involve strategic aggregation of two or more benefits so that benefits exceed costs. Benefit aggregation is necessary because individual benefits, even in "high value" circumstances, are rarely as high as MES cost. Similarly, the need to aggregate numerous individual, often year-specific benefits to cover the relatively high lifecycle costs of storage makes transportability quite attractive.

R&D Needs and Opportunities

The authors have identified the following activities as potentially fruitful research addressing attractive value propositions for energy storage, especially modular energy storage:

- Identify and characterize three to five emerging value propositions for MES characterized by specific criteria: 1) degree to which the value proposition is viable given: a) existing market mechanisms and b) expected and emerging market mechanisms, 2) ability to reduce regional blackouts (e.g. by proving local VARs), 3) expected utility infrastructure needs, 4) increasing penetration of intermittent renewables, and 5) increasing interest in "demand response" resources.
- Identify key technical and institutional challenges affecting the prospects for otherwise cost-effective use of MES by utilities, electricity end users, load aggregators and other third party electricity services providers, and characterize specific ways to reduce those challenges.
- Given results indicating that flywheel energy storage may be cost-effective as a transportable power quality resource, further investigation of that value proposition for flywheels is warranted.

TERMS USED IN THIS DOCUMENT

Demand – The rate of electric energy delivery, normally in units of kilowatts (kW) or megawatts (MW) for utilities (not adjusted for power factor).

Direct Cost – All direct costs to own or to rent an option, possibly including some or all of the following: rental charges, equipment purchase and delivery cost, project design, installation, depreciation, interest, dividends, taxes, service, consumables, fees and permits, and insurance. Direct cost reflects "point estimates" of future values, without regard to uncertainty.

Distribution – See Electricity Distribution

Electricity Distribution – Electricity distribution is part of the electricity grid that delivers electricity to end-users. It is connected to the transmission system which, in turn, is connected to the electric supply system (generators). Relative to electricity transmission, the distribution system is used to send relatively small amounts of electricity over relatively short distances. In the U.S., distribution system operating voltages generally range from several hundred volts to 50kV (50,000 Volts). Typical power transfer capacities range from a few tens of MWs for substation transformers to tens of kilowatts for small circuits.

Electricity Subtransmission – As the name implies, subtransmission transfers smaller amounts of electricity, at lower operating voltages than transmission. For the purposes of this study, "transmission and distribution" is assumed to include subtransmission and not high capacity/high voltage transmission systems.

Electricity Transmission – Electricity transmission is the "backbone" of the electricity grid. Transmission wires, transformers, and control systems transfer electricity from supply sources (generation or electricity storage) to utility distribution systems. Relative to electricity distribution systems, the transmission system is used to send large amounts of electricity over relatively long distances. In the U.S., transmission system operating voltages generally range from 200 kV (200,000 Volts) to 500 kV. Transmission systems typically transfer the equivalent of 200 to 500 megawatts of power. Most transmission systems use alternating current though some larger, longer transmission corridors employ high voltage direct current.

Equipment Rating – The amount of power that can be delivered under specified conditions. The most basic rating is the "nameplate" rating: nominal power delivery rate under "design conditions." Other ratings may be used as well. For example, T&D equipment often has what is commonly called an "emergency" rating. That is the sustainable power delivery rate under "emergency conditions" such as when load exceeds nameplate rating by several percentage points. Operation at emergency rating is assumed to occur infrequently, if ever.

Peak Demand – The maximum power draw on a power delivery system, usually year-specific. Subtransmission – See Electricity Subtransmission

Value Proposition – A value proposition is comprised of all benefits and all costs, including risk, that are associated with an investment or purchase.

REFERENCES

[1] Eyer, James M., Iannucci, Joseph J., and Corey, Garth P., "Energy Storage Benefits and Market Analysis Handbook," SAND2004-6 177, 2004.

[2] California Energy Commission and the Public Interest Energy Research Program, Electric Energy Storage Demonstration Projects in California, Request for Proposals (RFP) #500-03- 501. Attachment 14: Electric Energy Storage Benefits and Market Analysis, 2003. http://www.energy.ca.gov/contracts/RFP_500-03-501-07-31_RFP_500-03-501.PDF.

[3] Schoenung, S.M. and Hassenzahl, W.V, "Long- vs. Short-term Energy Storage Technologies Analysis: A Life-Cycle Cost Study," SAND2003-2783, 2003.

[4] Schoenung, S.M. and Hassenzahl, W.V, "Long vs. Short-term Energy Storage: Sensitivity Analysis," SAND2007-4253, 2007.

[5] Boyes, John D., "Comparing Costs with Benefits of Energy Storage Systems," EESAT 2003, October 2003.

[6] Schoenung, Susan M. and Eyer, James M., "Benefit and Cost Comparisons of Energy Storage Technologies for Three Emerging Value Propositions," EESAT 2005, October 2005.

[7] Mears, Dan, and Gotschall, Harold, "Economic Valuation of Energy Storage for Utility-Scale Applications," presented at ESA 2006, Knoxville, TN, May 2006.

[8] Lim, Janice (VRB Power Systems), personal communication. Also, VRB website: www.vrbpower.com.

[9] Buiel, Edward, "Axion Power International, e3 Supercell," presented at ESA 2006, Knoxville, TN, May 2006.

[10] Hadley, S.W., et al, "Quantitative Assessment of Distributed Energy Resource Benefits," ORNL/TM-2003/20, May 2003.

[11] PG&E Tariff Sheet, California Public Utilities Commission Sheet Number 23450-E, 2005.

[12] Biennial Assessment of the Fifth Power Plan, Gas Turbine Power Plant Planning Assumptions. The Northwest Power and Conservation Council. October 17, 2006. Page 2. http://www.nwcouncil.org/energy/Biennial/BiennialGasturbine.pdf

INDEX

A

accounting, 100
acid, 81, 83, 91, 92, 97
adjustment, 55, 97
aggregation, 7, 24, 59, 105
air emissions, 5, 53
algorithm, 72
alternating current, 3, 31, 64, 75, 107
American Electric Power, 3, 73, 75
APC, 93
arbitrage, 47, 80, 84, 89, 90, 98, 99, 104, 105
assessment, 46
assets, 73
authority, 59
average costs, 20

B

banks, 50, 71
base, 59
batteries, 22, 35, 73, 74, 75, 83, 91, 97
benchmarks, 58
beneficiaries, 63
benefits, vii, 4, 7, 8, 12, 21, 24, 30, 31, 34, 36, 42, 43, 44, 46, 47, 53, 54, 59, 61, 62, 63, 64, 65, 67, 74, 80, 82, 83, 84, 87, 88, 96, 97, 100, 102, 103, 104, 105, 107
biodiesel, 40
breakdown, 95
bromine, 81
building blocks, 36
business environment, 55
businesses, 56

C

cadmium, 81
candidates, 25
carbon, 91, 92
cash, 63
cash flow, 63
CEC, 68
challenges, 5, 24, 46, 59, 60, 61, 71, 72, 106
climate, 21
combustion, 5, 100
commercial, 2, 28, 29, 62, 80, 102
commodity, 47, 56
community, 72
compatibility, 46
compensation, 50, 71
competition, 59
complement, 43, 53
computer, 51
conditioning, 3, 31, 34, 46, 54, 71
conflict, 44, 50, 51
construction, 4, 24, 28, 29, 57, 66, 71
contingency, 63
cooling, 35

Index

coordination, 51
cost saving, 74
customers, 52, 71, 102
cycles, 33, 35, 73, 89
cycling, 72, 97

D

database, 18
deficit, 48
degradation, 35, 45, 105
demand curve, 39
Department of Energy, vii, 1, 2, 3, 68, 73, 79, 80, 81, 82, 83
depreciation, 63, 106
depth, 31, 35
direct cost, 24, 63, 106
discharges, 34, 35, 48, 97, 102, 105
disposition, 57
distributed energy resource, 3, 8, 23, 36, 46, 81, 82
distributed generation, 3, 5, 8, 24, 25, 49, 60, 63, 81
distributed utility, 3, 22
Distributed Utility Associates, 3, 81, 84
distribution, vii, 2, 4, 8, 17, 18, 19, 21, 22, 23, 26, 29, 33, 41, 46, 49, 63, 64, 71, 73, 75, 76, 77, 80, 81, 82, 84, 88, 106, 107
diversity, 23, 45
draft, 69
dynamic control, 72

E

economics, 74
Electric Power Research Institute, 3, 21, 67
electricity, vii, 2, 3, 4, 8, 9, 23, 24, 29, 30, 36, 47, 48, 49, 50, 52, 53, 54, 56, 59, 62, 63, 64, 66, 71, 73, 75, 76, 80, 82, 83, 84, 89, 99, 102, 103, 106, 107
electrolyte, 35
electronic circuits, 33
emergency, 51, 63, 64, 107
emerging technologies, vii, 4

emission, 39
employees, 2, 79
end-users, 54, 63, 65, 106
energy density, 36
energy efficiency, 8, 65
energy prices, 47, 52, 105
Energy Storage Systems (ESS) Program, vii, 82, 83
engineering, 29, 37, 83
environment, 71, 72
environmental effects, 59
environmental impact, 29
equity, 10, 62, 64

F

Federal Energy Regulatory Commission, 3, 17
financial, vii, 5, 11, 15, 23, 40, 49, 50, 55, 56, 62, 63, 64, 65, 74, 82, 83, 84, 85, 94, 96, 97
flexibility, 36, 55, 56, 59, 60
formula, 41
fuel cell, 22, 71, 91, 92

G

geology, 20
goods and services, 65
grids, 60, 75, 76
growth, 4, 7, 8, 11, 13, 21, 24, 26, 27, 28, 37, 38, 41, 49, 57, 65, 72, 74
growth rate, 8, 28

H

hot spots, 23, 26, 27, 28, 54
housing, 29, 61, 62
hybrid, 60

Index

I

improvements, vii, 4, 72
income, 10, 64
income tax, 10, 64
inflation, 16, 85, 86
infrastructure, 106
integration, 73
interest rates, 10
investment, 8, 23, 26, 36, 56, 58, 61, 66, 67, 74, 88, 107
investments, 8, 17, 29, 36, 56, 58, 61
issues, 90

J

Japan, 73
jurisdiction, 46

L

lead, 28, 29, 71, 75, 76, 81, 83, 95, 97, 101, 103, 104
lead-acid battery, 95, 101, 103, 104
leisure, 65
lifetime, 97
lithium, 81
low risk, 28

M

machinery, 33, 48, 66
magnitude, 7, 23, 28, 31, 47, 75
management, vii, 22, 24, 25, 48, 54, 65, 66, 80, 82, 83, 84, 89
marginal costs, 19, 20, 22, 23
marketplace, 53
Maryland, 17
mass, 33, 64, 65
matter, 46
meter, 71
Mexico, 1, 79
modular electricity storage (MES), vii, 2, 3, 4, 5, 7, 8, 12, 13, 14, 15, 16, 17, 23, 24, 26, 27, 28, 30, 31, 32, 33, 34, 35, 36, 37, 40, 41, 43, 44, 45, 46, 47, 48, 50, 51, 52, 54, 55, 58, 59, 60, 61, 62, 65, 69, 74, 80, 81, 83, 84, 88, 90, 91, 93, 94, 95, 97, 98, 104, 105, 106
modules, 65

N

NAS, 73, 74, 75
natural gas, 85
negative effects, 51, 72
nickel, 81
nodes, 26, 46, 49, 63

O

Oak Ridge National Laboratories, 3, 17, 67
operations, 57, 59, 61, 72
opportunities, vii, 4, 7, 31, 55, 56, 60, 66, 88, 99
optimization, 24
ownership, 57

P

Pacific, 3, 21, 81, 102
participants, 74
percentile, 22
permission, 37, 52, 59
personal communication, 108
physics, 34
polarity, 31
policy, 12
policymakers, 83
population, 20
population density, 20
portfolio, 29, 58, 60
power conditioning unit, 3, 31, 71
power generation, 50
power plants, 33, 49, 100, 105
precedent, 47, 52

present value, 16, 65, 74, 96
price signals, 104
principles, 59
probability, 28
profit, 55
project, 11, 13, 16, 24, 25, 46, 56, 57, 58, 63, 67, 73, 85, 86, 87, 106
property taxes, 64
proposition, 2, 4, 15, 24, 34, 43, 61, 62, 67, 83, 88, 89, 93, 94, 97, 98, 100, 102, 103, 104, 105, 106, 107

Q

quality improvement, 97

R

ramp, 33
Rate Assistance Program, 4, 19, 67
rate of return, 74
ratepayers, 10, 11, 13, 23, 58, 67
reactions, 40
real estate, 57
real time, 53
recall, 19
regulations, 5, 24, 40
reliability, 16, 24, 31, 43, 45, 51, 52, 57, 58, 60, 61, 67, 71, 72, 74, 75, 77, 89, 90, 94, 97, 98, 104, 105
renewable energy, 24, 49, 53, 54
rent, 56, 57, 63, 106
repair, 34
reputation, 24
requirements, 36, 39, 62
researchers, 83
reserves, 48, 105
resistance, 81
resources, 7, 8, 12, 23, 24, 25, 27, 28, 29, 36, 39, 42, 46, 50, 55, 56, 57, 59, 60, 61, 65, 82, 105, 106
response, 3, 8, 36, 49, 54, 55, 57, 59, 60, 63, 72, 105, 106
restrictions, 39, 56

retail, 47, 52, 56
revenue, 4, 10, 11, 13, 14, 15, 24, 26, 55, 62, 66, 67
rights, 2, 80
risk, 23, 36, 39, 46, 53, 55, 56, 58, 59, 60, 61, 67, 88, 107
risk management, 59
rules, 37, 58

S

safety, 5, 24, 61, 74
Sandia National Laboratories (SNL), vii, 1, 2, 4, 30, 37, 68, 69, 73, 74, 75, 79, 80, 81, 82
savings, 22
scope, 56, 60, 61, 83
security, 53
services, 45, 56, 57, 59, 106
shape, 7
signals, 46, 63
sine wave, 31
small communities, 71
sodium, 3, 35, 73, 74, 81
solid state, 33
solution, 49, 50, 55
stability, 49, 50
staffing, 34
stakeholders, 60
standard deviation, 19
state, vii, 2, 4, 73, 74, 80, 82, 93, 95, 101
stockholders, 10
storage media, 35, 36
structure, 52
sulfur, 3, 35, 73, 81
superconducting materials, 49
suppliers, 93

T

target, 17, 102
tariff, 74
taxes, 10, 63, 106
techniques, 58

technologies, vii, 4, 50, 83, 89, 92, 97, 101, 103, 105
technology, vii, 29, 35, 60, 72, 73, 74, 82, 83, 101
temperature, 36
territory, 23, 74
testing, 61
total energy, 39
trade, 2, 80
transactions, 47, 53
transmission, vii, 2, 4, 8, 17, 19, 24, 48, 49, 50, 54, 63, 64, 66, 73, 74, 80, 81, 82, 84, 88, 105, 106, 107
transmission and distribution (T&D), vii, 2, 4, 8, 80, 82
transport, 12
treatment, 59

U

unit cost, 25
united, 1, 2, 3, 31, 79, 80, 81
United States, 1, 2, 3, 31, 79, 80, 81
utility costs, 29

V

valve, 81
vanadium, 71, 81
variations, 21, 75
vehicles, 60
volatility, 53, 99
VRB Power Systems, 4, 71, 75, 108

W

water, 66
water evaporation, 66
wear, 31, 45
web, 68
wholesale, 47, 52, 56, 99, 105
wires, 19, 33, 60, 64, 107

Y

yield, 30, 88

Z

zinc, 81